A Chanticleer Press Edition

Albert C. Jensen

WILDLIFE OF THE OCEANS

HARRY N. ABRAMS, INC., PUBLISHERS, NEW YORK
Distributed in the U.K. by WINDWARD BOOKS

1–9. *The law of survival—"Kill or be killed"—is dramatized by killer whales (*Orcinus orca*), largest of the dolphins, and a sea lion (*Otaria flavescens*) in the surf off a lonely beach in Argentina. As the killers churn the water, the sea lion attempts to escape, but the struggle inevitably ends with the sea lion gripped in a whale's jaws. The kill is made, and the drama ends . . . for the moment.*

Library of Congress Catalogue Card Number: 78–27460
Jensen, Albert C.
Wildlife of the oceans.
A Chanticleer Press edition.
Includes index.
1. Marine biology.   I. Title.
QH91.J46   591.9'2   78–27460
ISBN 0-8109-1758-0

WHS Distributors
Euston Street, Freemen's Common
Aylestone Road
Leicester LE27SS

Composition by Neil W. Kelley, Middleton, Massachusetts. Color reproductions, printing and binding by Amilcare Pizzi, S.p.A., Milan, Italy.

Prepared and produced by Chanticleer Press, Inc., New York:
Publisher: Paul Steiner
Editor-in-Chief: Milton Rugoff
Managing Editor: Gudrun Buettner
Project Editor: Mary Suffudy
Production: Helga Lose, Ray Patient
Art Associates: Carol Nehring, Dolores Santoliquido, Johann Wechter
Editorial Assistant: Mary Beth Brewer
Picture Librarian: Joan Lynch
     Assistant: Jill Farmer
Maps: Richard Edes Harrison
Drawings: George Kelvin
Back Matter Drawings: Howard Friedman
Design: Massimo Vignelli

Illustration sources:
Pages 196–198
Sverdrup, Johnson, Fleming, *The Oceans*, © 1942, renewed 1970, pp. 289, 296, 815. Redrawn by permission of Prentice-Hall, Inc., Englewood Cliffs, New Jersey.
199–205, 207–209, 214–215
From *General Zoology* by Storer and Usinger, © 1957 by McGraw-Hill, Inc. Used with permission of McGraw-Hill Book Company.
206, 212–213, 220–221
After Davis, *Principles of Oceanography*, 2nd Ed., © 1977, Addison-Wesley, Reading, Massachusetts. Figs. 17.5 and 18.6. Redrawn with permission.
220–221
Sir Alister Hardy, *The Open Sea, Part II*, © Sir Alister Hardy, 1959, Houghton Mifflin Company.

Note: Illustrations are numbered according to the pages on which they appear.

Contents

Foreword

This book presents only a sampling of the hundreds of thousands of organisms that live in the seas of the world, but I believe it covers the most interesting and colorful. In some instances, for the reader's convenience, I chose to confine the discussion of a particular group of animals to the one area where these creatures appear most prominently, even though they also appear elsewhere in the sea. For example, although whales and dolphins are found in almost all parts of the world ocean, I discuss them mainly in the chapter on the open sea, since they are often found in that seemingly deserted area. Descriptions of the physical aspects of the sea—that is, the movement of waves, tides, currents, salinity, temperature, pressure, and especially the shape of the sea floor and crustal movements—are included here only as they explain why marine organisms occur in certain areas. Like land animals, marine animals are distributed in relationship to such physical features and conditions. Another criterion for selection was the availability of accurate information about an organism. Thus I concentrated on verifiable facts about the sea and its wildlife and engaged in speculation only when it had a reliable basis. The reader who seeks stories about sharks or killer whales that search out humans for revenge must look elsewhere. This book treats marine animals as they really are—organisms wonderfully adapted to their watery environment—and not as they are presented in fantasies.

And now a caveat for the reader. If you look to this book for visions of a future when mankind will reap riches from the sea, you will be disappointed. Yes, there is wealth in the sea, both in food and minerals; but the true wealth lies in better utilization of the sea's resources, especially its renewable resources. I have spent nearly a quarter of a century on the surface and, occasionally, in the depths of the sea, studying the world ocean in all its moods. As a result, I echo the pleas of many dedicated individuals who share my concern for the sea and its creatures. We must learn to conserve the sea and use it wisely. If we do not, our shortsighted exploitation will be our sorry legacy to generations yet unborn.

I am deeply grateful to the many people who helped me make the selection of ideas and illustrations for the book. Special thanks go to the staff of Chanticleer Press, particularly Milton Rugoff, Editor-in-Chief, and Mary Suffudy, Project Editor. I am also grateful to Loretta Miller for typing most of the manuscript.

Albert C. Jensen

Prologue

The sea is mysterious—a friend, an enemy, a provider, and, in a way, the genetrix of mankind. Indeed, each of us carries a "sea" in the salty blood that courses through our bodies. Although we now walk upright on the earth, we have a deep-seated relationship to the rhythmic cradle of our origin.

Our fascination with the sea and its creatures has been likened to an unconscious memory of an ancestral home. Even those among the human population who dwell hundreds of kilometers from the restless surge of the sea seem to experience a desire to go back to it; and sailors often speak of their love for the sea. But it is an unrequited love. She—and mankind has always so designated the sea—is ruthless, and we love her at our peril.

Our unending enchantment with the sea has led to intensive efforts to discover its origin. Studies have shown that more than 86 percent of the water on Earth is in the sea, while about 12 percent is bound up in sedimentary deposits and rocks, and a mere 0.03 percent of our planet's water is contained in lakes and rivers.

The Salty Seas

The massive bodies of saltwater represent an ancient feature of the Earth's surface. They have been present for at least 3 billion years, or nearly as long as the Earth, which is now thought to have been formed about 4.5 billion years ago. Geologists believe the seas were formed by the slow release of water to the surface during the eruption of volcanoes. Careful studies of marine organisms alive today and those which lived some 600 million years ago suggest that the composition of seawater has not changed very much over the millennia. The very characteristic of seawater, its saltiness, has been a source of much speculation among marine scientists. Folklore explains the salt in the seas as coming from a salt mill which was tossed into the original freshwater sea and which perpetually grinds out salt, making the water briny. Oceanographers say the sea was indeed a freshwater lake but became salty over the course of millions of years. Each of the rivers and streams emptying into the sea carries a small amount of salt dissolved from the rocks and soils in its watershed; it is believed that these have resulted in the salt sea we know today.

Ocean Basins and Drifting Continents

The various geologic forces that shaped the surface of the primordial earth formed the mountains, broad plains, continental shelves, and deep ocean basins. But the shapes of these features were not always as they appear today. At one time, the continents were all part of a single landmass that over the eons separated and ultimately formed the global configuration so familiar to every schoolchild. This theory of "continental drift," a startling concept, was set forth in 1912 by the German geologist Alfred Wegener. He noted that the bulging east coast of South America fits like a piece of a jigsaw puzzle into the deeply incurving west coast of Africa. And North America and Europe would fit together if the mass of Greenland were wedged between them to fill the remaining void. Wegener also thought that the land formations of Antarctica, Australia, and the southern tips of South America and Africa could dovetail. According to the theory of continental drift, the lighter materials of the continents float over the heavier and denser basaltic rocks of the Earth's core.

The concept of drifting continents did not fully satisfy marine geologists who were studying cataclysmic activity deep in the ocean basins. A more convincing, detailed explanation was formulated in 1968, by the French-American geologist Xavier Le Pichon, after he had carefully examined the activity of landmasses and the ocean floor. He theorized that the Earth consists of six large plates which are in continuous but imperceptible—and unmeasurable—motion. These plates constantly jostle one another, sometimes overriding adjacent plates, sometimes pulling apart. Their substantial movements are felt as earthquakes. Le Pichon named the plates and described the areas they affected, as follows:

Eurasian Plate: All of Europe, most of Asia, the East Indies, and the Philippine Islands.
American Plate: North and South America and the western Atlantic Ocean.
Pacific Plate: Most of the Pacific Ocean and small parts of North America's west coast.
African Plate: Africa, including Madagascar, the eastern half of the South Atlantic Ocean, and the western half of the Indian Ocean.
Indian Plate: The ocean floor from the Arabian peninsula to New Zealand, parts of southern Asia, and the island masses of Australia and New Guinea.
Antarctic Plate: The continental mass of Antarctica and small sections of the extreme southern Pacific Ocean.

Le Pichon's theory, now called *plate tectonics*, helped answer many questions that had long challenged marine geologists. The concept of shifting plates explained, for example, how new ocean floor is formed when plates move apart, thus allowing lava from the Earth's molten core to ooze out between the plates and congeal. It also explained the formation of such undersea mountain ranges as the Mid-Atlantic Ridge, which rises to great heights from the ocean floor where two plates move apart.

On land, the compression of plates creates great mountain ranges such as the Himalayas, Alps, and Andes.

Along some of the boundaries between oceans and continents, the moving plates not only cause mountain-building processes but are the sites of frequent earthquakes and active volcanoes. One of the most restless of these areas is along the shores of the Pacific Ocean and forms what is often called the "Pacific Ring of Fire." Awesome volcanoes and earthquakes that give rise to tsunamis, or tidal waves, and the mountain-building going on far beneath the ocean water all have shaped the ocean basins and given the seas the characteristics we know today. These events, occurring over countless eons, have also set the stage for the emergence of life in the sea and the evolution of marine animals.

14–15. *An undersea mountain range with great peaks winds 60,000 kilometers across the basins of the world ocean from the Arctic Ocean through the Atlantic and eastward into the Indian Ocean. Steep-sided seamounts and flat-topped guyots dot the submarine landscape. Molten material (magma) wells up from the interior of the earth through a rift in the range. Pushing outward, the magma causes the phenomenon called "sea-floor spreading."*

The age-old call of the sea is epitomized by a sunset over Kotzebue Sound, Alaska, near the Arctic Circle.

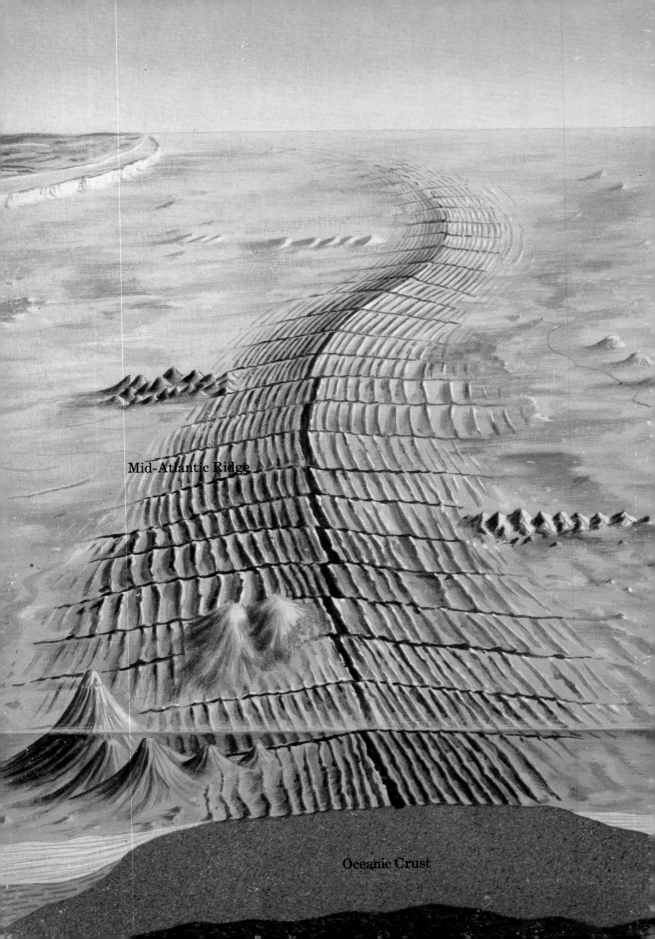

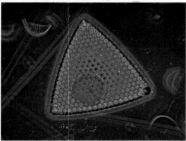

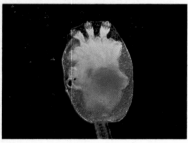

The tiny plants (phytoplankton),
which are the primary producers
of food and energy in the ocean
community, are eaten by the
small animals called
zooplankton. Top and second from
top. *The phytoplankton*
(diatoms) are plants; second from
bottom, the actinula (larval
coelenterate) and, bottom, the
tadpole of a sea squirt (larval
ascidian) are animals.
17. *Food webs in the sea are*
varied and complex. Here, the
process begins when the sun's
energy is transformed into food by
the phytoplankton, and is
completed as the swordfish eats the
squid.
18–19. *Among the top predators*
in the food chains of the ocean
ecosystem are the pinnipeds, such
as this group of California sea
lions (Zalophus californianus).

The Sea and Evolution

Most biologists agree that life began in the sea nearly three billion years ago—about as long ago as the time when the sea was formed. The first organism probably was a molecule that was able to duplicate itself, that is, to reproduce. Over the eons, more complex forms evolved, much like the plankton we see today.

With the passage of time, some plankton evolved into still more complex organisms, including those which branched off to create the groups of large plants such as rockweeds, kelps, and sea mosses, while others remained virtually unchanged. Similarly, some evolved into worms, sponges, corals, and, eventually, fishes.

A few marine organisms left the sea and evolved into land plants and animals. One group, after a sojourn of a few million years on land, returned to the sea and became the great whales and dolphins. But locked in the genes of these animals which are once again so much at home in the sea is the physiological memory of their ancient land-dwelling ancestors. This memory is routinely duplicated when the embryonic whale or dolphin rests in the little "sea" in its mother's womb. There, for a brief period, early in the gestation cycle, a pair of fleshy buds sprout on its tiny form. These buds, located where the hind legs of land mammals would be, disappear as the embryo grows. Occasionally, however, something goes awry, and one is born with identifiable (but nonfunctioning) legs protruding from its loins. A humpback whale processed by a west-coast Canadian whaling station was found to have hind legs nearly 1 meter long. Photographs showed the legs to be symmetrical, and dissection revealed bones and cartilage much like those in our own legs. And every whale and dolphin has within its flippers wrist and finger bones that look identical to those of land mammals.

Some marine mammals, including seals and walruses, presumably have not completed the evolutionary journey back to the sea and visit the land to rest and reproduce. One creature, the polar bear, has comparatively recently begun a return to the sea from which its ancestors departed so long ago. Even now it exhibits bodily changes that enable it to live in its harsh domain of an ice-girt sea. Its broad, well-padded, hairy paws, large compared with those of other species of bears, enable it to walk across the snowy expanses and noiselessly creep up on the seals it preys upon. Its broad feet also function very efficiently as swimming organs, propelling the great beasts many kilometers through arctic and subarctic waters. Perhaps many centuries in the future it will forsake the land entirely and return to its ancestral marine habitat.

The Ocean Community

The great numbers and variety of marine organisms have given rise to what is sometimes called a community, or *ecosystem*. The whole system is based on the simple phyto-plankton that use the nutrient salts in the sea and, with sunlight for energy, make food for themselves in the form of sugars and starches.

Because these organisms have little or no means of moving

themselves through the water, they are simply carried about passively by waves and currents. Plankton (a term derived from the Greek word meaning "to wander") are classified by scientists as plants, the *phytoplankton;* or as animals, the *zooplankton.*

The variety of drifting phytoplankton in the world ocean is virtually limitless but scientists have been able to group them into three main categories. These are: nanoplankton, diatoms, and dinoflagellates. The nanoplankton are the smallest of the three, often measuring less than 0.001 millimeter, and are the most important in the sea in terms of primary food production. Diatoms, on the other hand, are slightly larger. Under the microscope, they reveal an intricate, unique beauty. The outer skeleton is made of silicone and consists of two parts that fit together like a box and its lid. The "box" is elaborately sculptured with minute spines and holes arranged in complex designs. Sea water passing through the holes enables the tiny plants to carry on the life processes necessary for them to manufacture the food substances that in turn provide sustenance for larger organisms.

Dinoflagellates may be thought of as the Dr. Jekyll and Mr. Hyde group of the phytoplankton. Some of them create displays of great beauty. Others secrete poisons that are among the most powerful toxins known to man. The poisons kill birds and fishes and cause severe illness in humans. Benign species of dinoflagellates contain a substance called luciferin that combines with oxygen to produce a glow—bioluminescence. The organisms glow when they are stimulated by any physical, chemical, or mechanical agitation. The stimulus might simply be a wavelet moving through the water or, most commonly, the passage of a ship or the swimming movements of a fish, dolphin, or a man. The vast majority of dinoflagellates as with other members of the phytoplankton group are in fact harmless, beneficial members of complex food webs. Nearly all of the abundant life in the sea is dependent upon the "meadows" of microscopic plants for the basic supply of food.

The ecosystem, though complex, can be represented by the simple concept of a food chain, or web. Phytoplankton, for example, are eaten by zooplankton, and both kinds of plankton are eaten by clams, periwinkles, and sardines. In turn, still-larger organisms, including cod, tunas, sharks, and porpoises, feed on the clams, periwinkles, and sardines. Cod are eaten by sharks, which may then be eaten by other sharks. The result is that primal energy from the sun is passed as food from species to species, along very complex pathways. The whole ecosystem is a delicate balance of production and consumption, in which nothing goes to waste.

The ultimate consumers, of course, are members of the human species, who are apt to take too much and give as little as possible in return—to the detriment, and possible eventual destruction, of the ocean ecosystem. The future of that vital ecosystem now rests with mankind. Though land-dwelling humans have long stood in awe of the sea, they have now become responsible for the ecological health and survival of the ocean community.

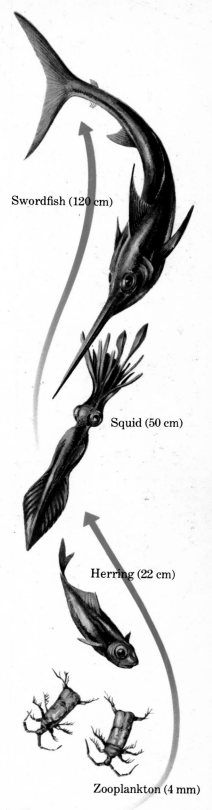

Swordfish (120 cm)

Squid (50 cm)

Herring (22 cm)

Zooplankton (4 mm)

Phytoplankton (1 mm)

The Seven Seas

For centuries, men have commonly referred to the "Seven Seas." We recognize today that in reality there are many seas, or only one—the world ocean. A glance at a map shows that the salt waters making up 71 percent of the earth's surface are continuous and joined to one another. The continents we call home are merely islands jutting above the surface of what might better be called "Planet Ocean" rather than "Planet Earth."

The oneness of the seas is demonstrated in the freedom with which mighty whales travel from one portion of the globe to another. Great fishes such as the blue shark, swordfish, and bluefin tuna also swim at will through the seas, with no regard to any artificial boundaries men have imposed. And many smaller forms of sea life, such as the invertebrates known as barnacles, mussels, and clams, are cosmopolitan in range, being carried from sea to sea on the tides and currents or on the iron skins of ships and the feathered skins of birds. The soft-shell clam (*Mya arenaria*) dug for supper by a bather on a shore of the North Sea is the same species as one that might be dug along the Gulf of Maine.

For convenience, we may say that there are three major oceans and that other bodies of salt water are just extensions of these three. The largest, the Pacific Ocean, encompasses over 165 million square kilometers. Next is the Atlantic Ocean, with an area of 82 million square kilometers. The smallest of the three, the Indian Ocean, has an area of 73 million square kilometers. Each of the major oceans extends into the frigid polar regions of the world to form ice-covered oceans. The Pacific and Atlantic reach the Arctic in the north and the Antarctic in the south, while the Indian Ocean extends only into the Antarctic.

The physical environments of the three oceans differ from one another, especially in the biologically limiting factors of temperature and salinity. (Salinity is defined as the total amount of dissolved salts in a kilogram of seawater. It is expressed as parts per thousand, that is, grams of salt per kilogram of seawater, and is designated by the symbol ‰.) Differences, especially those between the Atlantic Ocean and the Pacific Ocean, result from the shapes of the oceans. The Atlantic has been described as a twisted basin. Beginning in the frozen wastes of the Arctic and sub-Arctic, this basin broadens out to contain the waters that lap the shores of North America and Europe. It is broadest in the temperate zone between New York and the Straits of Gibraltar and narrowest along the equator. In the south, the Atlantic is broadest roughly between Buenos Aires and Capetown, another temperate zone. Thus, the greatest surface area of the Atlantic Ocean is in regions of moderate temperature, whereas its narrowest surface area is in the zone of tropical temperatures. The average temperature of the Atlantic is 10.5°C, and its average salinity is 35.3‰.

The Pacific Ocean, in contrast, has a nearly circular basin and is broadest at the equator and narrowest near the polar regions. Despite the vast amount of heat from the sun in the Equatorial Zone of the Pacific, its waters are only slightly warmer (11.0°C) than those of the Atlantic but are less salty (34.8‰). Similarly, the Indian Ocean, which has an egg-shaped basin, lies mostly in the tropics yet tends to be the coolest (10.0°C) of the three and has a salinity of 34.9‰.

The Seas

Even greater differences are found in those parts of the world ocean we have come to call the seas. The Baltic Sea is an excellent example, since its foggy and often storm-racked waters range from brackish to near-fresh in some parts but are quite saline in others, especially in the depths. The Baltic, actually an extension of the North Sea, is a long narrow basin between the Scandinavian Peninsula and the European mainland. With a surface area of 422,000 square kilometers, it has been described as the world's largest brackish water basin.

The salinity of the Baltic varies considerably, both in place and over time. At its entrance, where the oceanic circulation brings an inflow of quite salty water from the North Sea, the salinity is about 15‰. To the north and east, in the gulfs of Bothnia and Finland, where the inflow from the North Sea is diminished and is replaced by freshwater runoff from the land, the salinity drops markedly to only about 2‰. Salinity also varies seasonally and is decreased during periods of heavy rain or when the winter accumulation of snow and ice begins to melt in late spring and summer.

Just as the degree of salinity drops as one proceeds from the entrance of the Baltic to the interior, the number of species of marine plants and animals decreases. The most striking change is in the number of species of polychaetes, the marine segmented worms. Near the entrance to the sea, marine biologists have reported 160 species; in the middle Baltic, the so-called Belt Sea, the number declines to 70; in the upper Baltic, it drops to only 15; and in the gulfs of Bothnia and Finland, only 4 species of polychaetes have been reported.

Archaeological studies of the Baltic Sea reveal that it was not always an arm of the North Sea. During the Pleistocene—the Ice Age—the Baltic changed many times. It was sometimes covered by a crushing mantle of ice a kilometer or more thick. At such times, it may have been only an ice-filled basin that was further excavated and polished by glaciers during the eons of their slow advance and retreat. At other times, when the glaciers had melted northward into the high Arctic, the Baltic was an inland lake or was connected to the ancient Arctic Ocean. Fossil beds excavated in the Baltic Basin in recent times contain remains of the extinct arctic bivalve (*Portlandia arctica*) and the Greenland seal (*Phoca groenlandica*). A few "living fossils" can still be found in the brackish to fresh waters of the Baltic and the gulfs of Bothnia and Finland. These include the sculpin (*Cottus quadricornis*), a small freshwater fish found in polar waters around the world; and two small invertebrates, a mysid (*Mysis oculata*) and an amphipod (*Pontoporeia affinis*). While the Baltic Sea sometimes resembles a freshwater sea because of its reduced salinity, the Mediterranean tends

Top. *Hutton Cliffs, on Ross Island, between South America and Antarctica, are representative of the water expanses bound up in the permanently frozen ice fields and glaciers of northern and southern polar regions.* Bottom. *Icebergs, floating in the sea off Portage Glacier (Alaska) are remnants of massive chunks of ice that have broken off the glacier as it flows slowly to the sea.*

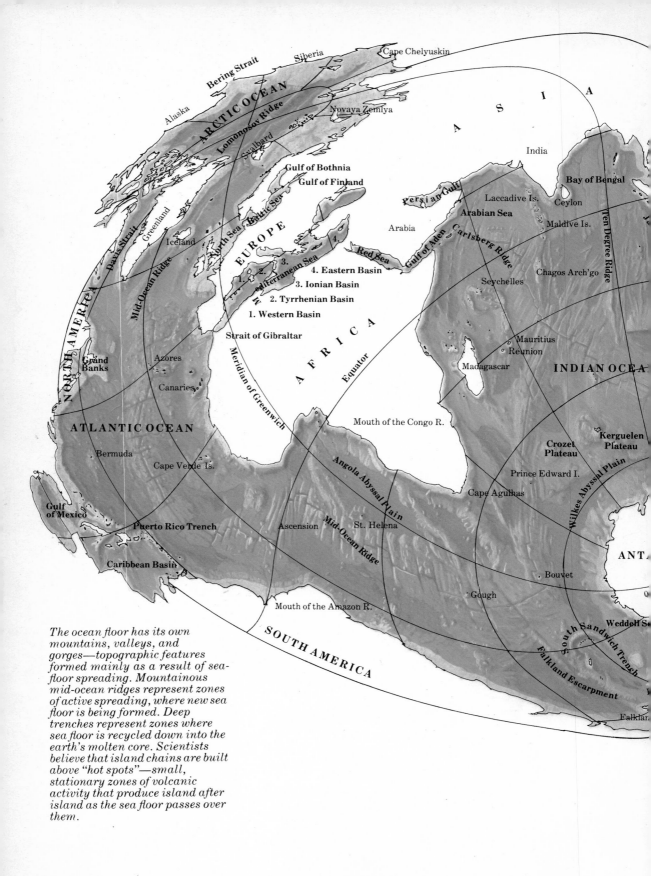

ARCTIC OCEAN

Bering Strait

Siberia

Cape Chelyuskin

Alaska

Lomonosov Ridge

Novaya Zemlya

Svalbard

A S I A

Greenland

Iceland

North Sea

Baltic Sea

Gulf of Bothnia

Gulf of Finland

India

Laccadive Is.

Bay of Bengal

Persian Gulf

Arabian Sea

Ceylon

Davis Strait

EUROPE

Mediterranean Sea

4.

Arabia

Red Sea

Gulf of Aden

Maldive Is.

Ten Degree Ridge

Carlsberg Ridge

Chagos Arch'go

Mid-Ocean Ridge

NORTH AMERICA

3.

1. 2.

4. Eastern Basin

3. Ionian Basin

2. Tyrrhenian Basin

1. Western Basin

Strait of Gibraltar

Seychelles

Mauritius

Reunion

Grand
Banks

Azores

A F R I C A

Equator

Madagascar

INDIAN OCEA

Canaries

Meridian of Greenwich

ATLANTIC OCEAN

Crozet
Plateau

Kerguelen
Plateau

Bermuda

Cape Verde Is.

Mouth of the Congo R.

Prince Edward I.

Wilkes Abyssal Plain

Gulf
of Mexico

Angola Abyssal Plain

Cape Agulhas

Puerto Rico Trench

Ascension

St. Helena

Mid-Ocean Ridge

ANT

Caribbean Basin

Bouvet

Gough

Weddell Se

Mouth of the Amazon R.

South Sandwich Trench

SOUTH AMERICA

Falkland Escarpment

Falklan

The ocean floor has its own
mountains, valleys, and
gorges—topographic features
formed mainly as a result of sea-
floor spreading. Mountainous
mid-ocean ridges represent zones
of active spreading, where new sea
floor is being formed. Deep
trenches represent zones where
sea floor is recycled down into the
earth's molten core. Scientists
believe that island chains are built
above "hot spots"—small,
stationary zones of volcanic
activity that produce island after
island as the sea floor passes over
them.

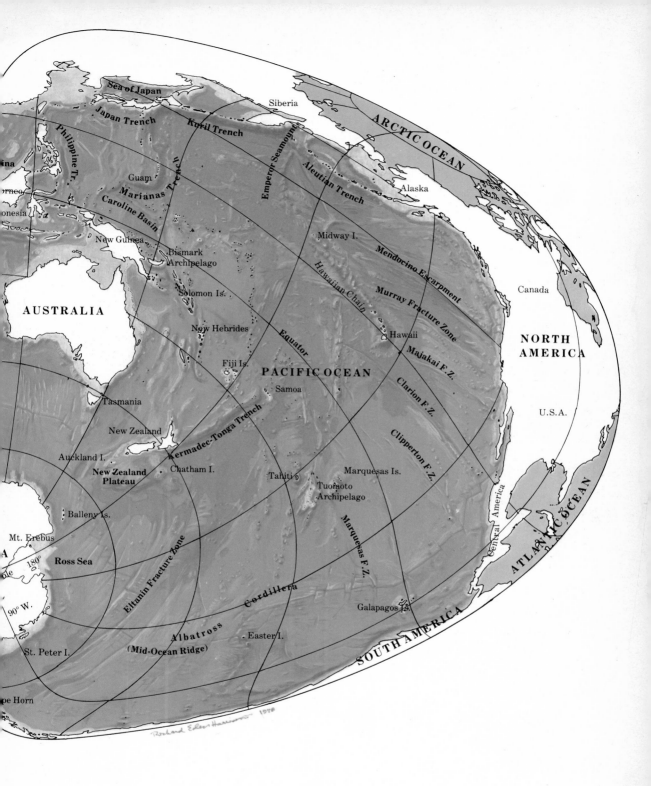

Sea of Japan

Japan Trench

Kuril Trench

Siberia

ARCTIC OCEAN

Philippine Tr.

Guam

Marianas

Caroline Basin

New Guinea

Bismark
Archipelago

Solomon Is.

New Hebrides

Fiji Is.

Emperor Seamount

Aleutian Trench

Alaska

Midway I.

Mendocino Escarpment

Hawaiian Chain

Murray Fracture Zone

Hawaii

Majakai F. Z.

Canada

NORTH
AMERICA

U.S.A.

Equator

PACIFIC OCEAN

Samoa

Clarion F. Z.

AUSTRALIA

Tasmania

New Zealand

Auckland I.

New Zealand
Plateau

Chatham I.

Kermadec-Tonga Trench

Tahiti

Marquesas Is.

Tuomoto
Archipelago

Clipperton F. Z.

Balleny Is.

Marquesas F. Z.

Central America

ATLANTIC OCEAN

Mt. Erebus

Ross Sea

Eltanin Fracture Zone

Cordillera

Galapagos Is.

SOUTH AMERICA

PACIFIC

180°

90° W.

Albatross

(Mid-Ocean Ridge)

Easter I.

St. Peter I.

pe Horn

Richard Edes Harrison 1978

China

Borneo

onesia

23

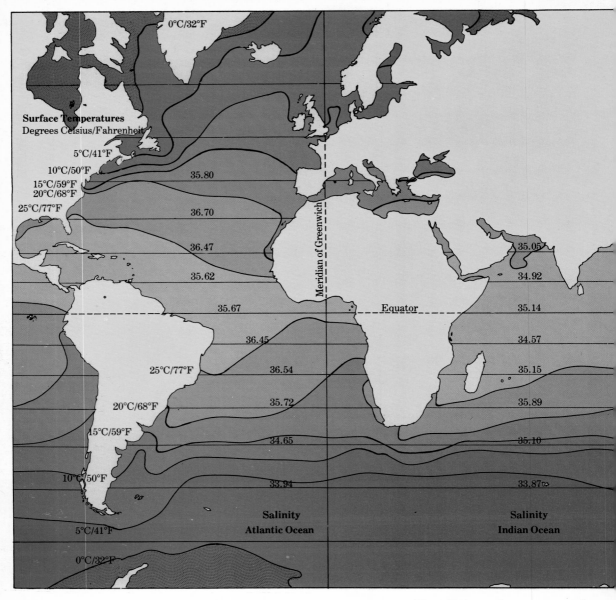

Surface Temperatures
Degrees Celsius/Fahrenheit

0°C/32°F

5°C/41°F
10°C/50°F
15°C/59°F
20°C/68°F
25°C/77°F

35.80

36.70

36.47

35.62

Meridian of Greenwich

35.05

34.92

35.67   Equator   35.14

36.45   34.57

25°C/77°F   36.54   35.15

20°C/68°F   35.72   35.89

34.65   35.10

15°C/59°F

33.94   33.87

10°C/50°F

Salinity   **Salinity**
Atlantic Ocean   **Indian Ocean**

5°C/41°F

0°C/32°F

The salinity and temperature of the world ocean vary from place to place. Rainfall, river flow, and evaporation from the water surface contribute to differences in salinity. The amount of sunlight falling on the ocean, the proximity of great ice masses, and the movement of water in ocean currents cause differences in temperature. The warmest water area in the world (29° Celsius) is indicated by the red patch on the map, just above Australia. The coldest areas of course occur mainly in Arctic and Antarctic waters.

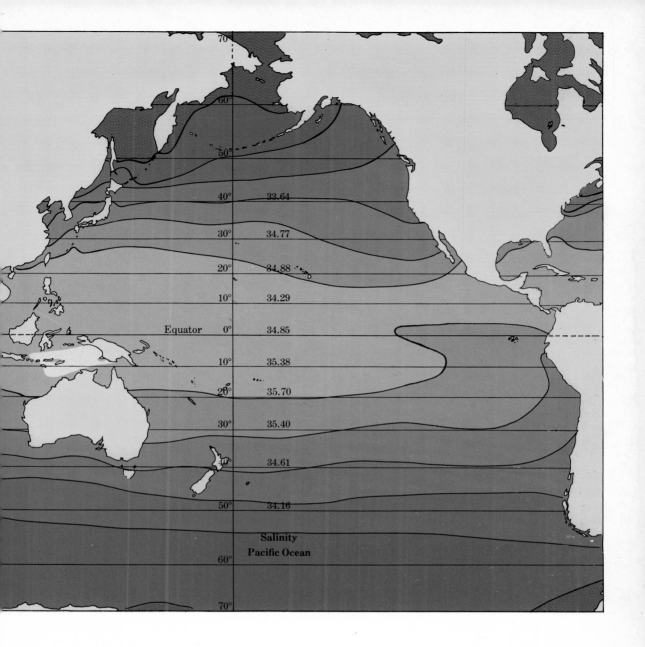

70°	
60°	
50°	
40°	33.64
30°	34.77
20°	34.88
10°	34.29
Equator 0°	34.85
10°	35.38
20°	35.70
30°	35.40
40°	34.61
50°	34.16
60°	
70°	

Salinity
Pacific Ocean

to be more saline than the Atlantic Ocean, with which it is connected. The submerged Gibraltar Ridge, which connects the Iberian Peninsula with the northwest coast of Africa, greatly restricts the inflow of cool, well-oxygenated water from the Atlantic. The warm and arid climate of the region around the Mediterranean causes its surface waters to evaporate at an increased rate, so that it becomes saltier than the Atlantic. At the eastern end of the Mediterranean, salinity reaches 39°/oo; some of this very salty Mediterranean water flows in a thin trickle over the sill into the Atlantic on the west, but most of it sinks to the depths of the sea.

Warm surface water from the Atlantic flows into the Mediterranean, so that the sea is warmer as well as saltier than the vast adjacent ocean. All the physical factors—the differences in temperature and salinity, the outward flow of saline waters over the sill, and the barrier created by the sill itself—serve to keep Atlantic deepwater species from entering the Mediterranean. Thus, most animals of this sea could well be characterized as surface-dwelling and pelagic.

Well-known members of the Mediterranean fauna are the dolphins, represented by two species: the common dolphin (*Delphinus delphis*) and the bottle-nosed dolphin (*Tursiops truncatus*). Leaping and swimming, riding the bow waves of ships, these playful mammals feed on the abundant fishes in the Mediterranean Sea. At times, their feeding activities interfere with commercial fisheries, because both man and dolphin often seek the same species, particularly sardines.

The most spectacular of Mediterranean fishes are the bluefin tuna (*Thunnus thynnus*), whose sleek, massive bodies—weighing 200 to 1,000 kilograms—move through the Straits of Gibraltar twice a year during their spawning migrations. In May the giant fish, well fattened after a nearly year-long feeding sojourn in the Atlantic, surge into the Mediterranean to seek their ancient spawning grounds. The females are plump, their bellies swollen with the millions of ripening eggs they carry and soon must release. By August, when the spawning ritual is completed, the tunas, now spent and seeking food, migrate westward through the straits from the relatively poor waters of the Mediterranean to the rich, productive waters of the Atlantic. During these enormous migrations the tuna are caught for food in great traps of stout netting set along the shores.

The sill across the Straits of Gibraltar is but one of the geologic features that determine the marine characteristics of the Mediterranean Sea. Other features of importance are the four deep basins that oceanographers have sounded; from west to east, these are the Western Basin, Tyrrhenian Basin, Ionian Basin, and Eastern Basin. The deepest spot in the Mediterranean is in the Ionian Basin, where Jacques-Yves Cousteau, aboard the *Calypso*, recorded a depth of nearly 5 kilometers.

On the other side of the world, the Sea of Japan reveals some features similar to those of the Mediterranean Sea.

The basin of the Sea of Japan is about 4,000 meters deep but is separated from the Pacific Ocean by sills that rise to

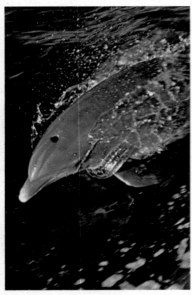

A bottlenosed dolphin (Tursiops truncatus) *surfaces to breathe. The 90 species of whales, dolphins, and porpoises belong to a group of mammals called cetaceans, which are all warm-blooded, air-breathing, and live-bearing. The nostrils (blowholes) are positioned on the highest part of the head and are directly connected to the lungs. These mammals do not breathe through their mouths.*

within 165 meters of the surface. During the glacial
and interglacial periods, the basin was elevated and then
depressed, and at times the sea became a broad lake
populated with freshwater fishes and other organisms. The
researches of modern zoogeographers have revealed
that a number of species of the Mediterranean and the Sea
of Japan are the same or are "twin species." The
experts believe these similarities have resulted from the
fact that the ancient Tethys Sea once was common to
the two regions. Millions of years ago, during the Eocene
and Oligocene, this ancient sea washed against shores
that are now parts of central and southeastern Europe
(including the modern Mediterranean Sea), northern
Africa, and western Asia (including the present-day Sea
of Japan). Today the Mediterranean and the Sea of Japan
are described as isolated seas—cut off, in a sense, from the
world ocean and characterized by oceanographic features
and living organisms not found in the world ocean.
But in this respect, both of these bodies of water are
exceptional; in general, the waters of the world ocean flow
into one another with considerable mixing of their fishes
and other organisms.

Life Zones of the Sea
Although the waters of the world ocean flow into one
another, as far as the life in them is concerned, they are
divided into "zones." One zoning system is horizontal,
with its subdivisions reaching from the shore out to
the open sea; the other is vertical and ranges from the
surface layers down to the ocean floor. These are not
compartments, of course; but they do affect—sometimes
quite dramatically—the kinds of life found in them.
If it were possible to walk on the surface of the ocean, an
observer could examine in detail the diversity and changes
in plants and animals from the shore out to the broad
stretches of the open sea. At the shoreline in a rocky
littoral zone, rockweeds (*Fucus*) and other large, attached
forms of algae sway with the ceaseless motion of the
waves. Tiny fishes dart through the shallow pools of water
between the rocks. In a sandy littoral zone, in contrast, few
if any algae are found, because the sand is too unstable
and shifts with each wavelet. A hardy alga might attach its
holdfast to a rock or shell to anchor itself on the sand.
The sandy littoral, however, is virtually devoid of
attached plants. Many larger fishes often boldly explore the
surf zone to feed on amphipods and other small invertebrates dislodged by the waves from their sanctuary
in the sand.
As an observer moves farther from shore, the number of
organisms he finds in the water increases. Herring shoals
move swiftly through the water, with mouths agape to
strain the abundant plankton in the nutrient-rich,
sun-drenched surface waters. Sharks, and occasionally
dolphins and whales, skim along the surface or break
through to feed on herring and mackerel. About 50 to 80
kilometers from shore the observer would see tunas,
swordfish (*Xiphias gladius*), and perhaps a few of the
billfishes. In warm temperate and tropical surface waters
the grotesque, disk-shaped ocean sunfish (*Mola mola*), 2 to

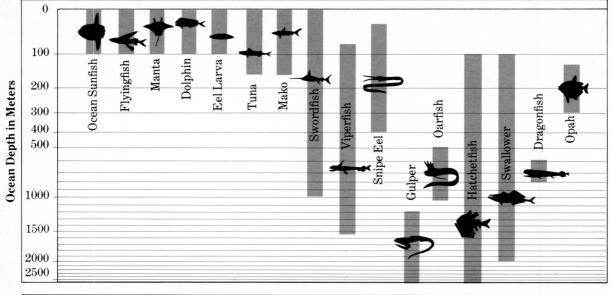

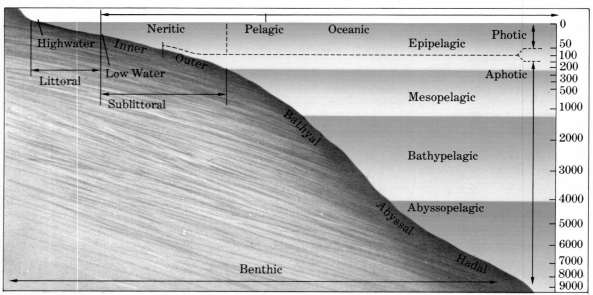

Top. *Pelagic fishes live in the open ocean. Most of them never encounter land—neither the shore nor the ocean floor. All of them are adapted for life within zones of water depth, either near the surface, at mid-depth, or deeper.*

Bottom. *Marine organisms tend to be either bottom-dwelling (benthic) or drifting-swimming (pelagic). The life zones that sea creatures occupy are arbitrary divisions based on depth and proximity to shore. In the littoral—or intertidal—zone, benthic and pelagic organisms live together.*

3 meters in length, swims clumsily along. Occasionally it pauses to feed on jellyfish and comb jellies (*Ctenophora*). Flying fishes dart from the surface and remain airborne, supported by broadened pectoral fins that act like the wings of a glider.

Far from land and the inflow of coastal rivers, upwelling longshore currents, and other sources of nutrients, mid-ocean resembles a biological desert. But life does subsist there, even though in reduced quantity. Then, as we approach the far side of the water, the opposite shore, the richness of ocean life increases until we again step onto the strand.

Horizontal distribution of plants and animals in the ocean depends largely on temperature, but vertical distribution —the second kind of life zone—of plants and animals depends primarily on light. Sunlight does not penetrate to all depths of the ocean, nor does it penetrate uniformly in all parts of the sea. The clarity of the water (hence the amount of sunlight that penetrates) is affected by solids and other substances suspended in the water column, including air bubbles, mineral and organic particles, and plankton. In the temperate and polar regions, abundant plankton greatly reduces light penetration.

Near coastal areas, river discharge and runoff from the land carry a load of soil particles that screens out the sun and reduces light penetration. In the tropics, where there is a paucity of plankton, sunlight penetrates to great depths in the clear, warm waters. Locally, however, light penetration may be reduced by solids carried out to sea in the tremendous volumes of water discharged by great river systems such as the Congo and Amazon.

The ocean floor closest to land is the continental shelf, averaging less than 130 meters deep and about 65 kilometers wide. The bottom topography of the shelf is similar to the adjacent land topography; that is, if the coastal area is rugged, so is the adjacent shelf, and if the coastal zone has plains or low hills, the shelf is likely to feature similar plains or low hills. This sublittoral zone covers only about 5 percent of the world ocean, but it is one of the richest zones for animal life. Nearly every group in the animal kingdom is well represented on the continental shelf. Ninety percent of the world's commercial fisheries are situated on the continental shelves, being concentrated at such famous fishing grounds as the Grand Banks, Georges Bank, and the North Sea.

Seaward of the shelf is the continental slope, a zone of rapidly increasing depths, from 130 meters down to about 2,000 meters. Animals of this zone are mostly dark and have very large eyes to enable them to see in what little light there is at such depths.

Beyond the continental slope, the continental rise begins a gentle descent to depths of 4,000 to 6,000 meters —the ocean basin. Beyond that lies the so-called abyssal plain, a zone at one time thought to be flat and feature-less; hence, a plain. However, the advent of echo sounders on ships made it possible to accurately survey the bottom and obtain profiles of this "plain." The readings made by the sounders revealed a complex and rugged topography. Volcanoes, most extinct but a few still active, dot the abyssal plain; many of them jut above the water surface to form familiar islands such as the Azores and the Hawaiian chain. Indeed, the latter group rises nearly 9,000 meters above the adjacent Pacific Ocean floor, or a height comparable to Mount Everest!

Seamounts, Guyots, and Trenches

Many oceanic peaks reach above the surface of the sea and form islands. Others do not rise so high, and these submarine peaks form seamounts and guyots. Seamounts have pointed tops, whereas guyots are flat-topped, as a result of once having been at sea level and then gradually eroded by wave action. All oceanic mountains—volcanoes, seamounts, and guyots—support life forms in distinct zones. From coral polyps near the surface to amphipods crawling over the ooze at the bottom, the organisms form a profile of the zonation of life characteristic of the world ocean.

The most spectacular features of the ocean basin are the trenches. Long, narrow, and steep-sided, these are the deepest depressions in the seabed. Most trenches are found in the Pacific Ocean, but a few occur in the Atlantic and Indian oceans. The deepest, the Marianas Trench in the southwest Pacific, plunges to 11,000 meters. In the deepest trenches the water temperature may be as low as −2°C, and the pressure as great as 30 tons per square meter. Despite great cold, tremendous pressure, and scarcity of food, animals live in these great rifts of the sea bottom. This was demonstrated conclusively on January 23, 1960, when Jacques Piccard, son of the famous Swiss scientist Auguste Piccard, and Lieutenant Donald Walsh of the U.S. Navy dived to the bottom of the Challenger Deep in the Marianas Trench aboard the bathyscaphe *Trieste*. They sank 10,912 meters, deeper than many scientists believed the ocean extended. From the *Trieste* the bottom appeared as a soft, cream-colored mud, and beyond the circle of illumination provided by the submersible's powerful searchlights there was only the unrelieved blackness of the abyss. But what was exciting was the sight of a flatfish, apparently a member of the sole family, about 30 centimeters long, that swam slowly out of the circle of light and into the eternal night of its domain.

The observations of Piccard and Walsh laid to rest at long last the ancient belief—cherished even by some competent scientists—of an *azooic* zone in the sea. Below a certain depth, in this zone, according to the theory, no life existed. At first, the azooic zone was thought to begin below the photic zone, that is, below the depth of light penetration. Then, as technology was improved and better nets, dredges, and other collecting devices were invented, and deep-sea cameras and underwater television permitted detailed viewing farther and farther down, the purported azooic zone was moved deeper and deeper. Finally, the various submersibles, such as the *Trieste* and *Alvin*, made possible direct observation of the bottom and actual collection of its rocks and animal life. It was thus proved conclusively that life existed in the ocean from the warm, sunlit surface down to the ice-cold black abyss.

31. *Formed through volcanic action, Iceland rests atop the Mid-Atlantic Ridge and grows as a result of continued volcanic activity. Thirty kilometers off Iceland's southern coast, in November 1963, a new island sprouted from the ocean. Youngest among the volcanic chain of the Westman Islands, which had long been inactive, it was named Surtsey. By 1970 nearly 160 species of insects, as well as a few land plants, had landed on Surtsey—as potential colonists from nearby islands or from the mainland.*

32–33. *Surf bathes the shoreline of a beach at East Hampton, New York. Since 71 percent of the Earth is covered by the salt waters contained in the three major oceans and their seas, it could perhaps have been more appropriately named "Planet Ocean."*

Great Currents and Tides

The sea is in a state of constant change. The most obvious change is its daily tides. For centuries, no one knew what caused the strange rise and fall of the sea that men called the tides. Many strange tales were fabricated to explain them. Some thought the tides were caused by a demigod who drained water from the ocean basin and later poured it back again. Others believed the tides were caused by the breathing of a giant whale. It was not until 1687 that the English scientist Sir Isaac Newton demonstrated that the tides result from the gravitational attraction of the moon and sun on the earth. Of the two, the moon's power is the greater. There are several aspects to the tidal cycle. First, there is the daily pattern, occurring approximately six hours apart, of the flood tide, when the waters are rising to the high-tide mark, and the ebb tide, when the waters are declining to the low-water mark. Then there are the spring tides (which have nothing to do with the season), when extremely high and low tides occur; and finally, the neap tides, when the tidal range is the least.

Flood and ebb are simply caused by the attraction of the ocean waters to the moon. Spring tides occur when the gravitational pulls of the sun and the moon are aligned during the time of the new and full moons. Some spring tides are so high they may cause flooding in low-lying coastal areas; their lows may be so extreme that they expose ancient roadbeds and building foundations long covered by the sea.

Neap tides occur when the sun and moon are at right angles to the earth during the first and third quarters of the moon. This periodic tidal range may be slight in areas where there is usually a rise and fall of only a half meter or so.

Tidal Marshes

The flooding and ebbing of the tide produces the tidal marsh, a coastal feature of many low-lying areas at the edge of the sea. Tidal marshes are among the most productive parts of the world, rich in a variety of plants, which in turn support an abundance and variety of marine animals. The tidal marsh is entirely different from freshwater marshes in a number of aspects—most of all its continually changing water levels resulting from the daily tides.

A unique feature of tidal marshes is the zonation of plants as one moves from inland toward the sea. At their highest point, farthest from the salt water, the dominant vegetation is a low-growing plant called black grass (*Juncus*), which is touched by salt spray only during storms. Closer to the sea, and covered by saltwater during the high spring tides, is salt-meadow cordgrass (*Spartina patens*). Nearest the sea, and covered daily by the normal high tide, is salt-marsh cordgrass (*Spartina alternifolia*).

Tidal marshes supporting cordgrasses can be found at the edge of the Arctic Ocean, along the coast of northern Europe, and around Australia and New Zealand. Extensive marshes occur along the east coast of the United States, even in such unlikely places as Jamaica Bay, on Long Island, near New York City, adjacent to Kennedy

International Airport. There, directly under the landings
and takeoffs of thundering jet aircraft, marine
organisms of all sorts carry out their lives.

In tropical and subtropical waters, tidal marshes are
dominated by trees rather than cordgrasses. The
edges of the coast support dense stands of mangrove
(*Rhizophora*), which has been called "the tree that walks to
the sea on stilts." Mangrove trees are very tolerant of
salt water, but their seedlings are not and thus are
protected from the killing salts by a unique growth pattern.
Instead of dropping from the parent tree to take root
and grow in the soil, the mangrove seeds develop into
seedlings while still attached. When the seedlings are about
25 centimeters long, they drop off and drift with the
currents until the root tip touches bottom and anchors in
the substrate. Here the familiar stilt-like prop roots of the
mangrove tree develop. As time passes, oysters become
attached and grow on the roots, while fishes and crabs move
among the roots to search out bits of food.

The abundantly rooted mangrove swamps, or marshes,
trap large quantities of silt swept in with the tide or
deposited by nearby rivers and, eventually, help form dry
land. When the mangrove is deprived of the salt water
it dies and leaves other trees to sprout and begin
the formation of forest. Extensive mangrove marshes
occur in the deltas of many great rivers such as the Mekong,
Amazon, Congo, and Ganges. Others develop along the
north coast of Australia and in the coastal areas of
Sumatra.

Mud flats are common to all tidal marshes. As the tide ebbs,
it exposes broad areas of seemingly barren mud or sand;
but the appearance is deceiving. From a few millimeters
below the surface of the mud flat, sometimes down to a
meter or more, the moist bottom supports a veritable
universe of marine animals, including clams, worms,
nematodes, and especially bacteria. All these animal forms
add to the overall productivity of the marsh.

Extreme Tides

The usual tidal range is determined by the size and
shape of the tidal basin. If the basin is wide, as in the
Mediterranean Sea, there is little rise and fall. If the basin
is narrow, as in the Bay of Fundy in Canada, the tidal
range may be great because a large volume of water is being
forced inland through a narrow space. The tidal range
in the Bay of Fundy, the greatest in the world ocean,
averages a difference of 15 meters between high tide and
low tide. When vessels dock in a port on the bay at high
tide, they must be firmly secured to the pier so they will
not tip over at low tide, when the water is too low to
support them.

Off the northwest coast of France, the little island of Mont-
Saint-Michel is daily isolated from the mainland by
extreme tides that, like those in the Bay of Fundy, may
have a range of 15 meters. Topped by a religious shrine of
world renown, the island is simply a large granite rock
thrust up from the seabed. When the tide begins to advance
across the surrounding exposed mud flats, it hisses and
rumbles in an onslaught against the tiny isle. In years

*Tropical birds, such as the roseate
spoonbill (Ajaia ajaja), nest by the
hundreds in mangrove trees
(Rhizophora). Mangroves are
tolerant of saltwater and help
protect tidal wetlands in the
tropics from erosion by hurricane-
whipped waves. The prop roots of
the mangrove have caused it to be
called "the tree that walks to the
sea on stilts."*

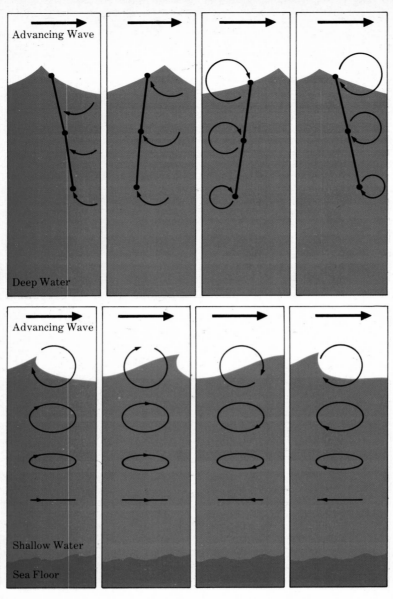

Advancing Wave

Deep Water

Advancing Wave

Shallow Water

Sea Floor

Left. *Contrary to popular belief, waves do not transport water forward; rather, the water particles (shown here as black dots) rotate in place. The orbit of particles decreases with water depth. In shallow water, the orbits become elliptical. The breaking of the waves at the shore, creating what we call the surf, or splash zone, is caused by friction between the bottom and the water particles when the waves enter shallow water. In a sense, the wave "trips over its own feet" and falls forward in a foaming crash. Air trapped in the wave crest forms the white froth of the breaker that dashes against the sandy or rocky shore. The water particles from the breaking waves then recede from the shore until they are picked up again by the energy of the next wave to break once more on the shore. And so the process repeats itself endlessly.*

37. The moon and the sun exert a gravitational pull on the oceans and cause the rise and fall of the tides. When the moon and sun are aligned, the highest, or spring, tides occur. When the moon and sun are at right angles to each other, the minimum, or neap, tides occur. The tidal cycle consists of a number of distinct phases. Flood tide begins as incoming water starts to rise and continues until the high tide mark is reached. At the midpoint of the flood cycle, the tidal current reaches its greatest incoming velocity. At the high tide mark the water movement is virtually nil— the high water slack. Ebb tide occurs when the water flow reverses and continues until the low tide mark is reached. At the midpoint of the ebb cycle, the tidal current again reaches its greatest velocity. Finally, at the low tide mark, low water slack occurs. Bathers and boaters have been marooned and some have drowned because they did not understand the range and speed of the tide in their locality.

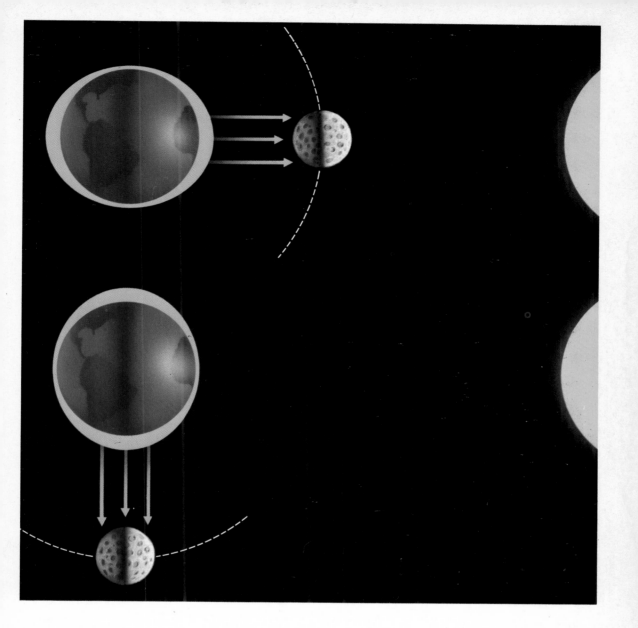

gone by, unwary pilgrims journeying across these flats to pray at the shrine were sometimes trapped by the incoming tide and drowned. Today a causeway offers safe passage over the quicksands and changing tidal waters. Fishermen take advantage of the extreme tides by setting out nets to snare shrimp swept in with the flooding waters.

Tidal Bores

Some rivers of the world have what are sometimes locally called "reversing waterfalls." These *tidal bores* are simply walls of swift-running water formed in a bay or river mouth by a rapidly rising tide. Tidal bores are most common where there are wide ranges of high and low tides. The incoming wall of water, which may be as much as 3 meters high, can cause small boats to founder.

The largest tidal bore in the world occurs in the Chang Tang Kiang River in China. The bore itself is nearly 3.3 meters high at spring tide and moves up the river as a wall of water at a speed of more than 25 kilometers per hour. It has been estimated that 1.75 million tons of water per minute is moved in this gigantic bore. Other well-known tidal bores occur in the Peticodiac River in New Brunswick, Canada; the Amazon River in South America; the Seine in France; and the Severn in England.

Tsunamis

The most devastating and damaging of the ocean's waves are the *tsunamis*, commonly called "tidal waves," although they have nothing to do with tides. These massive waves are caused by seismic disturbances such as volcanic eruptions or earthquakes. In 1883, the East Indian volcano Krakatoa erupted with great violence. The sound was heard for hundreds of kilometers, and ash from the eruption spread around the world. Its explosion also caused a violent tsunami that killed more than 36,000 people in the East Indies.

Most tsunamis occur in the Pacific Basin because it is surrounded by seismically active areas and volcanoes in what is sometimes called the "Pacific Ring of Fire." It is a curious aspect of tsunamis that they may travel thousands of kilometers across the open sea, yet be undetected by ships sailing over them. The wave length may be 200 kilometers, but the height may be only a half meter.

In water about 150 meters deep, the wave may move along at a rate of 75 knots (140 kilometers per hour), but in water 4,600 meters deep, it may speed along at a rate of 420 knots (780 kilometers per hour).

The damage occurs when the tsunami reaches shore, crests, and topples over. Then the wave becomes a towering watery mass wreaking destruction on homes, ships, and piers. The powerful surge of water may reach incredible heights: the tsunami from the Krakatoa explosion rose 35 meters above the beach when it touched shore.

Wind Waves

The most common kind of waves are those generated by the wind. These are the waves that make a boat bob and pitch. They come in a variety of heights and strengths.

As we stand at the rail of a ship at sea, the surface of the water seems to be moving in the direction of the waves, which in turn follow the direction of the wind. But the water really does not move; only the energy imparted by the wind moves in undulations across the face of the sea. When a rope is held at both ends and rhythmically flipped up and down, waves move along the rope, but the rope itself does not move forward. The same effect can be seen if a wood chip is dropped on the surface of water. The chip moves forward on each wave crest and backward in each wave trough, returning almost to its original position. The height of wind waves depends on the strength of the wind and the "fetch," or distance over the open sea that it blows. Some waves are small and barely noticeable because the wind is light; others, however, can be large, especially if strong winds blow for several days over vast expanses of the ocean. Although it is nearly impossible to measure wave heights accurately from a ship at sea, some heights can be estimated by simple triangulation. During a storm in the mid-Pacific in February, 1938, one of the officers of a naval vessel sighted a wave from the bridge through the top of the aftermast. By a simple trigonometric calculation, he estimated the wave to be 34 meters high!

The Beaufort Scale

Important to all who sail the seas is weather, and especially the wind. Recording weather at sea is a duty of all ships' crews. To help standardize the observing and recording of wind and sea conditions, in 1805 a British naval officer, Sir Francis Beaufort, devised a system by which certain sea conditions, especially wind velocities, could be recorded on a numbered scale. For example, Beaufort number 0 described a sea as smooth as a mirror, with the wind calm; number 5 described a sea with moderate waves, many whitecaps, some spray, and a wind of 17 to 20 knots, a fresh breeze. Number 12, dreaded by seamen, described a sea completely white with driving spray, the air filled with foam, and a wind of 64 to 71 knots—in other words, a hurricane. The Beaufort system of notation has been adopted by nearly every seafaring nation.

The Coriolis Effect

When the wind blows over the seas, surface waters or sea ice set in motion by the wind are deflected to the right of the wind in the Northern Hemisphere and to the left of the wind in the Southern Hemisphere. The oblique movements of the sea or ice are caused by the Earth's rotation to the east. This phenomenon is called the Coriolis effect, or force, after the French physicist Gaspard Gustave de Coriolis, who in 1835 explained the influence of the rotation of the Earth. The Coriolis effect explains the great slow-moving vortices set up in currents in the North and South Pacific, the North and South Atlantic, and the Indian Ocean.

Rivers in the Sea

The daily fluctuation of the tides is important to marine organisms, but there are other movements in the world

The Gulf Stream is visible as green in the lower portion of this satellite photograph of the east coast of North America. The Gulf Stream is the western part of the great clockwise current system of the North Atlantic Ocean. About 80 kilometers wide and 100 meters deep, it passes from the Gulf of Mexico through the Straits of Florida and then flows in a northeasterly direction along the coast of North America. Veering eastward at Cape Hatteras and moving at a speed of about 5 knots, it eventually reaches northern Europe. This warm ocean current, first described by Benjamin Franklin, is a moderating influence on the climate of the North Atlantic.

ocean, such as the great currents (called "rivers in the sea"), which are not easily seen but which have even greater impact on ocean wildlife.

In some parts of the world, ocean currents moderate the cold climates of lands that lie in northern latitudes, bringing comparative warmth to coastal areas of northerly locales such as Iceland, Norway, and Alaska. Other currents bring chilling damp and precipitation to lands such as western Canada and the U.S. Pacific Northwest. But all the currents bring food to the animals of the seas, transport the adults and their young, and frequently decide the fate of many of the sea's creatures. A well-studied example of such a current is the North Atlantic Current, best known—by the areas from which it originates—as the Gulf Stream.

The Gulf Stream–North Atlantic Current system flows at a rate of between 40 and 175 kilometers per day. In addition to transporting organisms that populate the seas of Europe, it carries many others to their deaths. Tropical forms such as the by-the-wind-sailor (*Velella*) or the pelagic snail (*Janthina*) perish as they are swept northward into cooler waters.

In the North Pacific, the Kuroshio Current is a warm, jet-like flow, similar to the Gulf Stream, that originates near the China Sea and flows generally eastward, eventually to become the North Pacific Current. As it approaches the west coast of North America, the current divides, with part of it flowing northward as the Alaska Current and part flowing southward as the California Current. Shrimp, salmon, and a host of bottom fishes thrive in the waters of the Alaska Current. Grotesque king crabs, some with legs measuring nearly 2 meters from tip to tip, prowl the chill bottom in search of food. Larger marine mammals, including the fur seal, walrus, polar bear, and, in season, the gray whale, seek their prey in these nutrient-rich waters.

To the south, the California Current carries water that is relatively cool and nurtures the great beds of brown algae, or kelp—forests of the sea—that bob and sway in the endless roll of the Pacific Ocean waves. There are several varieties of plants in the kelp forests, including *Laminaria*, *Nereocystis*, and *Alaria*. But the giant of them all is *Macrocystis pyrifera*, which may reach more than 30 to 35 meters from its root-like holdfast on the ocean floor to the tip of its frond, which functions like the leaves of earthbound plants in making food by photosynthesis. Joining the holdfast and the frond is the stipe, which might be likened to the trunk of a tree. *Macrocystis* specimens of up to 300 meters long have been reported along the California and Baja California coasts.

Small fishes dart in and out of the kelp forest like songbirds flitting through the branches of trees in a wood. Many-spined sea urchins creep along the stipes, browsing on the film of microorganisms and on the kelp, too. At times the urchins became so numerous they threatened the survival of the giant kelp. To keep the sea urchin population under control, biologists introduced sea otters—once hunted almost to extinction for their excellent fur—to feed on the voracious spiny

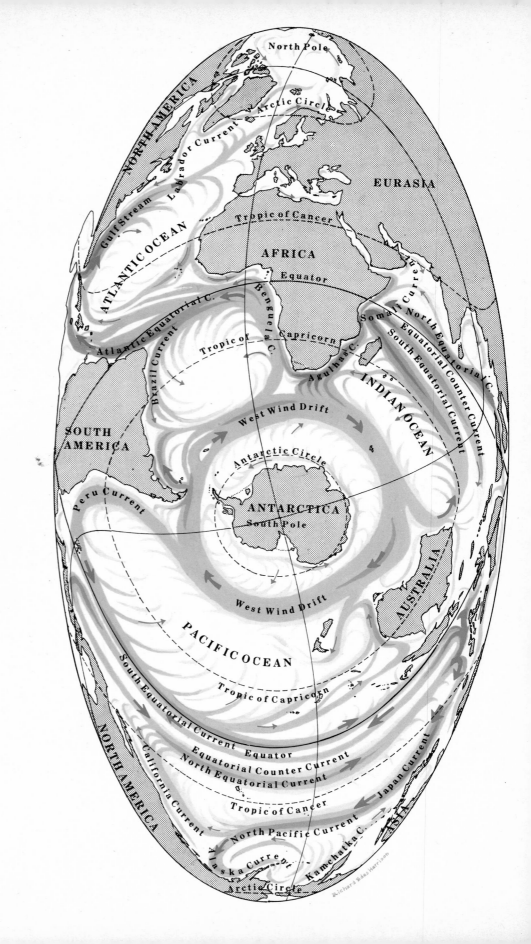

relatives of the sea stars, or starfish. Soon, balance of a sort was restored. The otters kept the urchin populations down to the point where their browsing and feeding on the kelp stipes had little effect on the density of the sea forest.

Sweeping eastward in an unbroken circle around the world, the West Wind Drift flows past the frozen reaches of the ice-locked south polar continent. With no landmasses to impede its path, this rough and stormy belt of frigid water streams between Antarctica and the southernmost reaches of the continents of Australia, South America, and Africa. As it moves, the West Wind Drift thrusts immense chilling streams into the lower portions of the South Pacific, South Atlantic, and Indian oceans. Two important intrusions of the current make contact with the west coasts of South America and Africa. These intrusions carry cold, well-oxygenated water and contribute to the richness, diversity, and productivity of the coastal and offshore waters of these mighty continents.

One intrusion, the Benguela Current, flows past Capetown, along the west coast of Africa, and then north and west-ward to cross the Gulf of Guinea. The cold portions of the current off Capetown nourish dense kelp forests near shore and contribute to the wealth of animal life in the cold waters over the narrow continental shelf, including the stockfish (*Merluccius capensis*); the maasbanker (*Trachurus trachurus*); and the dense schools of shoaling sardine, or pilchard (*Sardinops ocellata*). Yellow-tan spiny lobsters (*Jasus lalandii*) crawl over the sea bottom, probing with their long antennae for bits of food or safe refuge from enemies. These armored animals lack the large claws of the common lobster, and some purists insist they be called "crayfish" to differentiate them from their northern counterparts.

The other intrusion from the West Wind Drift is the Peru Current, also called the Humboldt Current, which sweeps northward along the west coast of South America. Named after the famed German explorer and geographer Alexander von Humboldt, it is the southern counterpart of the California Current but is colder because of the burden of cold Antarctic water it carries. The cold current and the upwelling associated with it bring to the surface nutrients which result in one of the most productive aggregations of animals in the world. Anchovetta (*Engraulis ringens*), silvery relatives of the sea herring, feed on the abundant plankton. They, in turn, are pursued and consumed by a host of predators, including tunas, whales and porpoises, and seabirds. For countless centuries, flocks of birds have plunged into the sea off Peru to gorge on the anchovetta. Then, sated with food, the birds fly to nearby rocky islands to rest. With the passage of time, the bird droppings left on rocks have formed guano deposits, hundreds of meters thick. Local fishermen reap a double harvest from the wealth delivered by the Humboldt Current. With great seines they encircle the shoals of anchovetta; the catches are then ground and pressed into meal and oil. The deposits of guano are mined and, along with exports of the meal and oil, are shipped as fertilizer for farms and gardens around the world.

Above. *The eye of a hurricane—a powerful tropical cyclone—is an area of calm at the center of a storm with winds in excess of 120 kilometers per hour. Originating over tropical seas, hurricanes (called typhoons in the western Pacific) may measure more than 400 kilometers in diameter.*
Right. *Storm-driven waves and winds up to 240 kilometers per hour batter the beaches of Miami, Florida, during a hurricane.*

But ocean currents can be fickle, and at times seemingly as capricious as the winds that blow over the land. At infrequent intervals, warm equatorial waters spread south and displace the cool waters of the Humboldt Current. The nutrient-rich Antarctic water is replaced by nutrient-poor tropical water, and a catastrophic cycle begins among the marine life. First the primary producers, the phytoplankton, die—and the fish that depend on the plankton, like so many cattle that graze in the pastures of the sea, die, too. Beaches become littered with windrows of dead fish, and fishermen find their nets nearly empty. The guano-producing birds suffer as well; the large flocks that grew fat and abundant during the years of plenty now starve, and their bodies also litter the beaches. In this relentless cycle, hunted and hunter perish alike.

This marine catastrophe occurs at about the Christmas season and is called *El Niño*, a Spanish name for the Christ Child. But this *Niño* brings no welcome gifts; indeed, the prospect of its coming is dreaded, because it not only causes wholesale destruction of animal life along the coast but means poverty and near-starvation for fishermen and their families. It is fortunate that this disruption of the cycle generated by the Humboldt Current does not happen with greater regularity. *El Niño* has occurred in 1925, 1930, 1941, 1951, 1953, 1957, 1965, and 1972.

The Sargasso Sea
Sometimes currents, or lack of currents, produce strange effects in the ocean. The Sargasso Sea, found in the central part of the Atlantic Ocean, has been the source of myths and legends for centuries and was once called the "Graveyard of Ships" because vessels were sometimes trapped there in its great masses of seaweed. The weed itself, a brown alga (*Sargassum*), is a tropical species that accumulates on an immense scale in the central Atlantic. Portuguese sailors who first saw the vast patches of weed likened the gas-filled floats on the plants to grapes (sarga), thus giving the sea its name.

The Sargasso Sea is the center of a current system that brings in the floating weed. But the currents in the middle of the sea are weak, and the entangling weed accumulates along with logs, parts of wrecked ships, and other flotsam. This is also an area of weak and variable winds, so that a sailing ship might lie there motionless for days while the sun beat down with fierce intensity. Although ships and men have been lost in the Sargasso Sea, it was not because of mysterious monsters, but rather because of the curious natural—but perverse—conditions in this part of the world ocean.

Hurricanes and Waterspouts
Literature abounds in stories of great storms at sea. Some storms are local and may do little or no damage. Others are regional, last several days, and do great damage. These last are the tropical cyclones—a general term for any circular windstorm—called

hurricanes in the Atlantic, and *typhoons* in the Pacific. They are vast disturbances with wind speeds in excess of 120 kilometers per hour, and some have been recorded with winds of up to 300 kilometers per hour. They may cover an area of more than 400 kilometers.

On the open seas, hurricanes and typhoons raise waves 15 meters or so high, which crest and break as if crashing against the shore. The violent winds tear at the top of the waves, flinging foam in horizontal sheets that, combined with rain, make it impossible for a sailor to distinguish between air and sea. Only a stout ship can weather a hurricane at sea.

Hurricanes and typhoons may signal their coming by periods of gathering clouds and increasing winds. But waterspouts rise spontaneously, often from calm seas during periods of bright sun and no wind. Such spouts are compact cyclonic storms that appear as a column of water over the sea and are kept upright by a whirling motion of the wind. They resemble a funnel that descends from a cloud bank and meets a cone of spray rising from the sea. The entire waterspout column may be from 3 meters to over 1 kilometer high and from 3 to 150 meters wide. Waterspouts are short-lived and last only about half an hour; they rarely cause any damage. When a spout breaks up, it releases a torrent of fresh and salt water that may be accompanied by sleet, snow, or hail.

Climatic Influences

Storms and waterspouts are extreme phenomena of weather, but the sea also exerts influence through climate, which is a relatively long-term process. Its influence on climate is a consequence of the oceans' serving as gigantic heat reservoirs: that is, the ocean water has great capacity for absorbing the heat from the sun and storing it. During the cold-weather months, the enormous quantity of absorbed heat is slowly released into the atmosphere and then to the land. For this reason, coastal lands are not so cold in winter nor so hot in summer as areas a few hundred kilometers inland. In some places the differences can be very marked. An extreme example is the monsoon climate of the Indian subcontinent. In summer the interior regions of India are heated by the sun. The heated air rises and is replaced by moisture-laden air drawn in from the Indian Ocean. This causes the Indian summers to be hot and humid. In winter the inland areas grow cold, while the adjacent ocean remains warm. The warmed air over the ocean rises, thereby drawing off cold air from the subcontinent's interior. These seasons and the accompanying winds are called the summer monsoon and the winter monsoon. The change from one season to the next is dramatic: air temperature can drop or rise as much as 20 degrees within a matter of hours.

Mankind has long dreamed of controlling the weather for such desirable aims as lessening the fury of hurricanes or increasing the warmth and growing season of agricultural lands. But the sea is too vast, and the climatic and weather systems it generates are too complex. The natural forces that induce the phenomena of weather and climate remain beyond our control.

47. *An awesome waterspout forms over tropical seas near Key West, Florida. Whirling columns of seawater, spray, and rain are contained in this powerful, tornado-like windstorm.*
48–49. *A brilliant orange sunset over the Gulf of Mexico silhouettes a brief but intense summer storm. Such storms are often accompanied by flashes of lightning and crashing thunder.*

The Littoral Zone: A Hostile Habitat

The first exploitation of the resources of the world ocean took place in the littoral zone, that part of the shore between the high-tide line and the low-tide line. Even today, the shore and the littoral zone are the only parts of the marine environment that most people see. There is a special pleasure in collecting the shells, driftwood, mermaid purses, and other materials cast up on the beach by waves. There is something soothing and restful in the gentle, ceaseless motion of wavelets on a calm, sunny day. There is awe and perhaps fear at the sight of a breaking surf in a storm. Finally, there is the fascination of ever-changing life revealed by the ebb and flow of the tides. This constant flux in the littoral makes life there very precarious for some plants and animals, yet very hospitable for others.

The Rhythm of the Tides

The daily rhythm of the tides is like some great clock regulating the lives of the inhabitants of the littoral. At low tide, the zone is exposed to the drying heat of the sun in summer and to the killing frost in winter. As the tide rises, drawn by the irresistible power of the moon, waves dash against the rocks, pebbles, and sand forming the shore. Winter storms tear at the littoral, scouring great hollows in the sand and sometimes hurling boulders many meters onto the land. At high tide, the littoral is flooded with a protective cover of seawater that shields the animals and plants from heat and cold. Wave action may be lessened at high tide, but in a severe storm the entire littoral becomes turbulent with the ceaseless pounding of great waves.

But this exchange of water means life to the organisms of the littoral. The waves enrich the water with oxygen and, more important, bring in food particles, loosen amphipods and other small animals, and stir up nutrients from the bottom. Barnacles, mussels, and limpets, firmly attached to the rocks, feed on the rich broth swept in by the rising tide. The barnacles extend their feeding baskets, rhythmically sweep them through the water, and trap microscopic organisms. Fishes and crabs forage among the rocks, and worms crawl over the bottom in an endless search for bits of organic matter.

The kinds of animals and their activities in this zone depend on whether the littoral is a rocky or sandy shore or is situated at the edge of a tidal marsh. Tidal marshes, rich with rooted plants and among the most biologically productive areas in the world, may very well have been the cradle of terrestrial animal life.

The Dawn of Animal Life

Animals had long been abundant in the ancient seas, but the land was untenanted except for various kinds of plants. Then animals appeared on the land, perhaps to escape the deadly competition of the marine world. Presumably they emerged in a tidal marsh. Although the sea teemed with predators, it did provide a relatively stable environment. The marsh, on the other hand, was subject to great and frequent changes; it was alternately flooded and drained by the ocean tides in an endless cycle. There eventually evolved in the marsh habitat great land animals that no longer depended upon the sea. Some animals remained in the marshes and evolved into forms well suited for life in that changing environment. Many of these animals can be studied today in the salt marshes that abound along coastlines bathed by the world ocean.

Marsh Life

The kinds of animals found in a typical tidal marsh depend on the part of the marsh in which they live. Around the edges, above the reach of all but the highest storm tides and waves, are the parade grounds of the bizarre-looking fiddler crab (*Uca*). Generally less than 3 centimeters long, fiddlers dig burrows in the moist marsh soil, going down as much as 60 centimeters. When undisturbed, they scuttle about at the edge of the tidal flat, in search of bits of organic matter for food. These little crabs expertly roll the food into tiny spheres, which they eat directly or carry down into their burrows for a later feast.

The male fiddler gives the species its name because one of its claws is astonishingly large and is carried as a violinist carries his instrument. The other claw is a more normal size and shape. The male makes elaborate motions with his fiddle when courting females or threatening enemies. The female has no "fiddle."

The muddy, sandy flats exposed at low tide contain considerable quantities of detritus, a mixture of sand or mud and plant and animal remains resulting from the action of the water. Many of the animals in this exposed zone feed on detritus, swallowing an almost continuous stream of the mixture, digesting nutrients from the organic portion and excreting the sand and mud together with undigested organic remains. Some animals are filter feeders; that is, they strain bits of detritus and plankton from the water when the flats are flooded at high tide. Other animals are carnivores that prey on their neighbors and sometimes on their own kind.

One of the most common cannibalistic predators of the tidal flats is the mud snail (*Nassarius vibex*). This gray-brown gastropod is only about 12 millimeters long, but its ferocity belies its small size. These snails locate food by scent and travel freely over the sandy flats in search of it. They devour clams or other bivalves by boring through the shells of their prey with a rasp-like "tongue."

They are also not above attacking other mud snails and may bore through the shell of a close neighbor in their voracious quest for food.

Another inhabitant of the flats is the razor clam (*Ensis*), a filter feeder. Resembling a closed, old-fashioned straight razor, this clam may be 50 to 70 millimeters long. It rests upright at the bottom of the low-water mark, with its posterior end and siphons extended. Wading birds, including gulls and herons, relish the razor clam and stalk across the tidal flats searching out the exposed portions of this bivalve. If disturbed, the clam slips down out of sight, rapidly drawing itself into its burrow by means of its strong, muscular foot. This clam is also a favorite of the predatory moon snail (*Polinices*), but has evolved a quite effective escape response; when threatened by its slow-

52–53. *Snails, numbering nearly 70,000 species and occurring everywhere in the sea, are mollusks of the class called Gastropoda. They are primarily adapted to life on the ocean bottom, but some swim and others float. Numerous species occur in fresh water, as well as on land. Snails are absent only from glaciers and mountain peaks.* Row 1: left, Turbo petholatus; center, *smooth turban* (Norrisia norrisii); right, *moon snail* (Lunatia heros). Row 2: left, *banded tulip* (Fasciolaria hunteria); center, *wavy topshell snail* (Astraea undosa); right, *African volute* (Volutidae). Row 3: left, *harp snail* (Harpa major); center, *turban snail* (Tegula); right, *Gray's volute* (Amoria grayi).

Fiddler crabs (Uca), *generally less than 3 centimeters long, are common in saltwater marshes all over the world. The male has one enlarged claw that is held like a violin: hence its familiar name. He stands near the entrance to his burrow and waves his claw at a passing female, who may then join him in his burrow, where mating occurs.*

moving predator, the razor clam slips out of its burrow and quickly begins to dig into a new burrow.

Changing with the Tides

Anyone who has watched the creatures that feed in tidal waters will see that birds are often among the predators. Some, such as gulls, feed at low tide, but others, such as terns, actually plunge into the water in pursuit of prey as the tide comes in.

The flats exposed at low tide invite birds that scurry about, picking over bits of fish left by the receding waters. Other birds hunt for worms, clams, and crabs. The most active and noisy are the gulls of the genus *Larus*, which squabble raucously over the most putrid fish carcass. The gulls also fiercely snatch at and fly off with any crabs or clams unfortunate enough to be exposed. To open bivalves, the gulls carry them aloft, hover over a rocky area or even over paved roads, and drop the shellfish onto the hard surface. If the shell does not break on its first fall, the gull retrieves it and repeats the ingenious process.

The surge and splash of the incoming tide enriches the water with oxygen and food particles and marks the start of feeding by the filterers. The rising water level also is a signal for the incursion of animals from below the tide line. Searching for food, small shrimps and several species of crabs move onto the now-submerged flats.

Fishes, too, make forays into the littoral zone. The robust-looking killifishes (*Fundulus*)—some about the size and shape of a man's thumb—dart about the bottom as they pick up tiny invertebrates or neatly snatch minute diatoms, planktonic plants carried in with the tide. In season, schools of juvenile bluefish (*Pomatomus saltatrix*) rush in like wolves to attack and devour the killifish. Rippling and undulating like flags held sideways in a breeze, young plaice (*Pleuronectes platessa*), in the eastern Atlantic, and young and adult winter flounders (*Pseudopleuronectes americanus*), in the western Atlantic, search through the littoral for food. Eagerly they dart at worms, amphipods, and other invertebrates exposed as they themselves feed on the bottom.

Graceful terns, dainty relatives of the gulls, wheel and dart in flight over the flooded marsh as they snatch small fishes and crustaceans from the waters. Sometimes hovering in midair on fluttering wings a few meters above the water surface, they scan the shallows for schools of silversides (*Menidia*) or killifish. As soon as potential prey is in range, the tern plunges into the water, then emerges instantly with the still-wriggling captive in its strong beak.

Attached Animals

In contrast to the rich life of the littoral zone of a tidal marsh, there seems to be a paucity or even absence of organic life in the littoral zone of a rocky shore. Indeed, at first glance, life would seem impossible in an environment that lacks the enriching support of the muddy tidal flats. Here, too, the land is periodically drenched by the waves, washed by rain, and dried by the sun. But there is life here—a compact, tenacious kind of life.

When the marbled godwit (Limosa fedoa) *probes for the worms, clams, and crustaceans that are its staple diet, its 12-centimeter beak sometimes is not long enough. Although generally considered a shorebird, the marbled godwit also breeds on the grassy plains of Canada and the northern United States, where it feeds itself and its chicks on insects.*

Above. *A western gull* (Larus occidentalis) *stands in a California tide pool with a starfish snatched from the water. These large birds, about 50–60 centimeters long, are common along marine coasts and estuaries. Western gulls usually forage for themselves on rocky shores, but they may also force cormorants and pelicans to release their prey, which the gulls then seize.*

Left. *The Caspian tern* (Hydropygne caspia), *which breeds along the coastal areas of every continent except South America, feeds on small fish; but it may also rob the nests of nearby seabirds, devouring their eggs and young.*

"Red crab," or lobster krill (Pleuroncodes planipes), *occur seasonally in the hundreds of billions off the coast of Baja California and southern Mexico. Usually found offshore within 4 meters of the surface, some are occasionally stranded on the beach.*

Twice in recorded history (1859, 1960) mass strandings occurred in Monterey Bay, California, some 1,200 kilometers north of the usual range of the species. The 1960 stranding, which formed a band 2 meters wide and 100 meters long, was estimated to contain 40,000 individuals, each about 7 centimeters long. Lobster krill, which are decapod crustaceans, are eaten by whales and fishes, particularly tunas of several species. Like lobsters, krill are equipped with claws (chelipeds)— a resemblance that was noted by the whalers who discovered and named them.

Left. *A lobster krill wriggles out of its old shell during the molting process. Like other crustaceans, this animal must shed its shell, or outer skeleton, in order to grow.*

Sea stars usually have five arms radiating from a central disk, inside which are digestive and reproductive organs. Some species evert the stomach through the mouth, on the lower side of the disk, and digest food outside the body. Clams and oysters are favored food items. Typical sea stars are, top, *batstar* (Patiria miniata); center, *cobalt sea star* (Linckia); *and* bottom, *pink sea star* (Pisaster brevispinus).

The animals of such zones have evolved through the eons to take full advantage of the rocks and the more meager food supply. With ingenious arrangements of form, size, mode of attachment, and manner of feeding, they have adapted well to the seeming inhospitality of their environment. The diminutive, flattened forms of starfishes (*Asterias*), limpets (*Patella*), and chitons (*Chiton* and *Chaetopleura*) greatly reduce the animals' exposure to the tearing and battering of the surf. To increase their ability to hold fast to rock, limpets erode a scar on the stony surface in which they implant their firm, muscular "foot." Thus anchored, the limpet can resist a pull of nearly 32 kilograms. Chitons, too, increase their ability to resist the pull of the waves by adhering to the rock with a sinewy foot; similarly, the starfishes cling with tube-feet and also secrete a glue-like mucus to further increase their adhesive power. But all these adaptations have their price. They may enable the animals to remain in place under the pounding surf of a raging coastal storm, but they also force them to live a slow-moving, nearly fixed existence.

The strongest defense against the relentless pull of the waves is made by the animals that attach themselves permanently to the rock. The white, conical shells of the barnacles (*Balanus, Chthamalus*) become nearly one with the rock by means of an adhesive cement that the animal secretes. Nothing can break the grip of this cement except the animal's death, and even then the plates of the empty shell may remain attached for years. The mussel (*Mytilus*) uses a complex of sturdy threads called a byssus to form a flexible but tough fastening. A special gland secretes a liquid that on contact with seawater hardens to create a thread. One thread at a time is secreted and attached to the rock until the animal is secure.

Each change in the tide changes the animal's environment. Half the time the mussel is exposed to air, at the mercy of the heat or cold and its drying action; during the other half the mollusk is submerged, busily filtering plankton from the turbulent surf-zone waves.

Although the starfish is unable to crack the strong defenses of the barnacles, it finds the bivalve mussel an easier prey. Gliding over the rock on its tiny, flexible tube-feet, the starfish probes for food. The motionless, helpless mussels, growing in profusion, present a ready meal. The starfish embraces a mussel and holds the two halves of its shell in an unyielding grip. Now begins a relentless tug-of-war that usually has but one outcome. With its arms, the starfish exerts a powerful pull, trying to force apart the mussel's shell. The pull is maintained steadily, for hours if necessary. Eventually the mussel tires and its adductors relax, causing the shell to gape helplessly. Now the victorious starfish everts its stomach and proceeds to digest the soft tissues of the mussel. Soon there is nothing left except the two empty valves still held fast to the rock by the now-useless byssal threads.

Algae Browsers

In contrast to the fixed and immobile animals, chitons cling tightly to the rocks but do make small excursions for

60–61. *At low tide, sun stars (Solaster dawsoni) cling to rocks on a beach along the Pacific Northwest coast of the United States. When the tide returns, the sun stars search out smaller sea stars, sea anemones, bivalve shellfish, and snails, which they eat in large numbers. In common with other sea stars,* S. dawsoni *show great variation in color.*

Below. *Limpets are small, snail-like mollusks with conical shells. They are abundant in the intertidal zone. Top. A two-spotted keyhole limpet* (Megategennus gimaculates) *clings tenaciously to a rock at Moss Beach, California. Bottom. The common European limpet* (Patella vulgata) *clamps tightly to the rocks when the tide is out but moves about to browse on seaweed when the water returns.*
63. *Chitons, 3-centimeter-long mollusks with overlapping plates that form their shells, cling to the walls of tide pools with such strength that they seem to have been bolted down. Top. Sea cradle* (Tonicella lineata); *center, T. lineata, showing color variation; bottom, coat-of-mail shell* (Polyplacophora).

food. The chitons are gastropods, but with a shell composed of eight overlapping plates and a broad, flat foot on the underside. Capable of creating a vacuum between itself and the rocky substrate, this animal can resist virtually any pull. Should the chiton become dislodged, it has one final defense. Tucking its head under its tail, it rolls itself up in a ball and tumbles away from predators.

A veritable Samson among the algae browsers is the limpet. The most common one is *Patella vulgata*, found on the rocky shores of northern Europe. These animals, which have squat shells that look like upside-down funnels, can cling to their precarious perches with an attachment that resists a pull equivalent to nearly 8 kilograms per square centimeter. From its home base, the limpet makes small excursions to glean the film of algae and diatoms that is nourished by the life-giving splash of the waves.

Periwinkle Life Zones

If one had to select an animal as most typical of the rocky shore, it would have to be the periwinkle (*Littorina*). Among the several species of this humble marine snail, four are common on European shores bathed by the temperate waters of the Atlantic. All of them browse on the algae film that mantles the rocks, but each has a preferred level on the rock in relation to the splash zone.

All periwinkles can survive extended periods of exposure during the low-tide period. Some may retreat to tide pools and there await the return of the surf. Others simply close their operculum, like a trapdoor, and remain attached to the rock, often in the baking heat of the tropical sun.

Flexible Anemones

Some attached animals of the rocky shore do not use adhesive strength to resist the power of the waves but, like a boxer rolling with his opponent's punches, absorb the force of the rising waters instead. Chief among these are the sea anemones, which look like plants rather than animals and present a pliant body to the surf. The sea anemone is simply a large polyp, similar to the smaller polyps that make up coral reefs. Each polyp consists of a stout, muscular stalk, from 1 to several centimeters long. At the foot end of the stalk is the disk by which the anemone attaches itself firmly to the rock. At the mouth end, where the stalk flares slightly, like the bell of a trumpet, are situated the numerous hollow tentacles that cause the sea anemone to resemble the flower whose name it bears. Most sea anemones are only 1 to 3 centimeters in diameter at the mouth end, although the warm, sun-drenched waters of the Great Barrier Reef of Australia nurture anemones measuring as much as 1 meter in diameter at the mouth end.

The brilliant white, green, blue, orange, red, and varied combinations of these colors that suffuse the body of the anemone, especially its tentacles, make it appear even more like a vivid blossom. But these colorful tentacles form no delicate flower; instead they are part of an efficient, often deadly feeding apparatus.

64–65. *Although named for the wildflowers of the mountains and woodlands, the familiar sea anemones of tide pools and rock ledges are animals, which defy description in their variety of shape, size, and color. Some are plankton and detritus feeders, while others are active predators, trapping and consuming small fishes. All these sea anemones belong to the family Aliciidae— except, top row left, Endomyaria.*

Above. *Fan worms* (Bispara) *extend plumose tentacles to sweep the water for plankton and other microscopic food particles. When disturbed, the animals quickly retract the fans into the tube in which they live. The worms measure 5 to 15 centimeters long.*

67. *Nudibranchs are shell-less snails. Along the back of the body are numerous projections (cerata), usually arranged in rows. Nudibranch means "naked gill," but no true gills occur in these animals.* Top. *A purple nudibranch* (Flabellinopsis iodinea) *glides over the bottom of a tide pool, its antennae probing for food or detecting enemies.* Bottom. *As* Chromodoris californiensis *demonstrates, brilliant coloring is characteristic of sea slugs.*

At low tide the sea anemone is unable to feed and so draws its tentacles down into the stalk and shrinks into a rubbery mass. At high tide, when it stretches its body, unfurls its tentacles, and moves them sinuously about its mouth, the anemone resembles the mythological Gorgon, Medusa. Now the tentacles, armed with paralyzing darts—nematocysts—become a snare to trap unwary fishes and small invertebrates. Should a small fish blunder into the anemone, it is quickly stunned by the nematocysts. Numb and helpless, it is then borne by the anemone's tentacles into the gastrovascular cavity and digested.

The worm-like tentacles of some sea anemones occasionally serve as feet when the animals move to a choicer spot on the rocks. Other anemones move by a slow, almost imperceptible gliding on their attachment disk. Sometimes an anemone will simply drift off with a water current. There is great danger in this last form of locomotion because a passing fish, such as a cod, may see the meaty morsel in the water and snap it up.

68–69. *The dorsal plumes, or cerata, of* Antiopella barbarensis *gleam as if illuminated from within, as this tropical sea slug moves through its watery domain.*

Marine Feather Dusters

Tide pools frequently are home for some of the most spectacular marine animals: the tube-dwelling, feather-duster worms. Related to the simple earthworm of field and garden, the feather-duster worm builds tubes that may be as much as 45 millimeters long. Shyly extending its body from the open end of the tube when the tide has flooded its rocky habitat, the worm deploys a symmetrical array of brightly banded gill filaments. This display resembles a gaily colored, old-fashioned feather duster; hence their common name.

A specimen of feather-duster worm common on the shores of Europe is the honeycomb worm (*Sabellaria alveolata*). It builds its tubes in masses that at first glance look like the honeycomb cells of beehives. At low tide the worm retreats into its tube and plugs the open end with its two front feet. It also snaps back into the tube at even the slightest vibration or when a shadow falls on the plumes. When the tide is at flood level, the filaments of the worm extend to a radius of about 7 or 8 centimeters, and tiny cilia on the filaments begin a rhythmic, wave-like beating. The beating cilia propel food particles into the worm's mouth.

Naked Shellfish

In many parts of the world ocean, occasional visitors to the littoral zone are members of a bizarre group of "shellfish" without shells, the nudibranchs, or sea slugs. Unlike other gastropods, such as the periwinkle, nudibranchs have no shell or, at best, a greatly reduced internal shell. Sea slugs look like a psychedelic explosion of colors. Indeed, they often are called "sea fairies" or "rainbows of the sea." Grotesque appendages jut from their multihued bodies. At the head end, sea slugs feature two sensitive antennae that constantly probe for food. The rest of the body—or sometimes just the hind end—bears myriad extensions that resemble multicolored trees, leaves, plumes, and even balloons.

One group of sea slugs feeds on hydroids and sea anemones. When it preys on the deadly tentacles of an anemone, the sea slug secretes huge amounts of mucus over its victim and then bites chunks off the prey. The nematocysts, or stinging cells, of the sea anemone are stored in appendages on the back of the sea slug, ready to explode and sting any fish that nibbles on this colorful marine booby trap.

The Cruel Antarctic Littoral

The harshest conditions in the littoral zone anywhere in the world are found on the edges of the Antarctic continent. Its interior is overlaid with an ice dome that may be as thick as 3,000 meters. During the brief Antarctic summer, when the ice melts and retreats from the shore, an inhospitable rocky littoral is exposed. A small variety of plants and invertebrates flourish for a few weeks, but life there is scarce indeed. However, this is the domain of the familiar dumpy, tuxedo-clad penguins. From the large emperor penguin (*Aptenodytes forsteri*), a stately 1.2 meters tall and definitely imperial, to the diminutive Adélie penguin (*Pygoscelis adeliae*), a modest half-meter tall, these flightless birds have come to symbolize the frozen continent of the South Pole.

Penguins remain on the frigid wastes year-round, some mating in the sunless winter and raising their one chick during the short summer. In winter the birds are buffeted by fierce storms, with winds raging up to 320 kilometers per hour and temperatures dropping to −85°C. In the summer, when either parent goes off on a feeding expedition, it is eagerly pursued in the sea by its dreaded predator, the leopard seal (*Hydrurga leptonyx*). The seal savagely attacks the penguin as it "flies" underwater with powerful strokes of its flipper-like wings. The unfortunate bird is no match for the seal, which quickly reduces the once-natty-looking penguin to a few bloody scraps of skin and feathers.

Nor are the penguin chicks secure from enemies, despite the close care of their parents. Skuas (*Catharacta skua*), or "sea gulls turned into hawks," attack the downy chicks and devour them. Other chicks are killed when elephant seals (*Mirounga leonina*) drag their ponderous bodies across penguin nurseries. Mature elephant seals may be as much as 7 meters long and weigh up to 3,300 kilograms, so their journey through a penguin nursery is marked by a trail of crushed chicks.

But predators are not the sole destroyers of penguins. Accidents among the ice floes take their toll, too.

For example, a sad fate eventually awaits several large colonies of birds adrift on an ice island 48 kilometers long, 42 kilometers wide, and 180 meters thick that broke off from the shelf ice of Antarctica in 1967, complete with its unknowing residents. As of 1978, this frozen island was drifting toward the southern tip of Africa, where it is expected to break up in the warmth of temperate and subtropical waters. During the 11 years of their odd journey, the floating island's penguins have produced many generations of birds. But once they reach the warmth of the African waters, being unable to endure the heat there, all will die.

A chin-strap penguin (Pygoscelis antarctica) *poses on a barren beach before a gleaming Antarctic ice cliff. Penguins feed almost exclusively on fish and other marine animals, which they capture by expert diving and swimming in the chill southernmost sea.*

Top. *Adélie penguins* (Pygoscelis adeliae) *plunge from an Antarctic ice shelf into the sea to begin a hunting foray. Although strong and graceful swimmers, penguins are awkward when they walk on land.*
Bottom. *A leopard seal* (Hydrurga leptonyx) *patrols an Antarctic ice floe, waiting for Adélie penguins to enter the water. Ferocious predators that even killer whales avoid, these seals are powerful swimmers, able to lunge out of the water onto ice 3 meters above.*

72–73. *A leopard seal catches and kills an Adélie penguin, then shakes the bird out of its skin and swallows it whole. This aptly named predator of the south polar seas, which may grow to 4 meters long, is exceeded in size only by the elephant seal.*

The Shallows Zone: Cradle of the Seas

The shallows zone adjacent to the landmasses of the continents is appropriately called the continental shelf. In some parts of the world, such as off the northeast coast of North America and in the North Sea, the continental shelf is wide (250 kilometers maximum). Elsewhere, such as off the west coasts of North and South America and Africa, the shelf is narrow (approximately 75 kilometers). But wide or narrow, the continental shelves are extremely productive zones that nurture most of the world's valuable fisheries. They are the natural bases of great and complex food pyramids that support a marvelous array of marine plants and animals—and, ultimately, man himself.

The shallows zone over the continental shelf, generally less than 200 meters deep, is the home of plaice, herring, cod, and other important food fishes. Crabs and lobsters scuttle over the bottom in search of food, while scallops and clams strain food particles from the currents that sweep across the shelves like wind across a pasture.

Submerged Forests

The continental shelves, of course, are clearly part of what we like to call "the land." In fact, an oceanographer has likened the great mass of a typical continent to a hippopotamus submerged in water with only its back and the top of its head exposed. The exposed parts are like dry land, while the great beast's shoulders, just beneath the water surface, form a continental shelf. The shallows zone over the continental shelves makes up only about 8 percent of the area of the world ocean but is equal to about 20 percent of the total land area of the world.

But the shelves have not always been submerged. In the Northern Hemisphere, only about 11,000 years ago, the continental shelves were exposed for an average distance of 115 kilometers from the present shoreline. Men once lived, hunted, raised families, and died on these now-submerged lands and left evidence of their ancient ways in the form of great mounds—kitchen middens—of oyster shells in what is now very deep water, many kilometers out from shore. Great beasts roamed the forests and meadows of these long-ago coastal areas. Even today, fishermen dragging their trawl nets along the floor of the continental shelf occasionally haul up the teeth, bones, and tusks of mammoths (*Mammuthus jefersoni*).

Early man used the seas first as a source of food rather than a means of transportation. Wading in the shallows, he fished with spears and crude maze-like traps fashioned from bushes and branches and, later, with nets woven from vines or thongs. Shellfish of all kinds were abundant, and no devices were needed to gather these, except for rocks to smash them open. Indeed, before the domestication of animals, coastal dwellers depended on fish and shellfish as a primary source of protein. Today, marine fish and shellfish annually contribute nearly 60 million tons of the world's diet. Of that amount, 78.5 percent is harvested from the shallows zone; about half consists of finfish caught with nets or hooks in the surface and near-surface waters, and about half are fish, mollusks, and crustaceans caught with hooks, dredges, and traps on or near the seabed.

Harvest of Abundance

The immense productivity of the shallows zone is the result of a combination of factors. There is an abundance of light throughout the water column to provide energy for the phytoplankton and attached algae. As previously explained, the foundation of all food production is the phytoplankton. Diatoms and other drifting green plants absorb the nutrients in the surface layers and, with energy from the sunlight that penetrates the heaving waves and swells of the sea, begin their complex chemical magic. Within the plant cells, the elements are transformed into sugars, starches, fats and vitamins. Each plant multiplies rapidly; mother cells cast off daughter cells, until the waters become what one oceanographer has described as "the broad pastures of the sea." Very quickly these marine grazers begin to feed on the rich bounty. The zooplankton, particularly the copepods such as *Calanus* and their relatives, feast on the lush, green oceanic pastures. The tiny animals gorge themselves on the abundant plants and very quickly begin their own population explosion.

In many parts of the world ocean, sea grasses, submerged flowering plants more complex than the simple algae, are the primary producers. Such flowering marine plants include eelgrass (*Zostera*), in temperate and cool waters, and turtle grass (*Thalassia*) and manatee grass (*Cymodocea*), in tropical waters.

These slender-leaved plants grow in lush, undersea meadows, usually in less than 1 or 2 meters of water. Their leaves sway to and fro in the wash of coastal inlets and lagoons. Unlike kelps and other attached algae, which anchor themselves with a holdfast, sea grasses are true rooted plants that require a soft substrate. Their growth helps to keep the soft bottom from being scoured by strong winds and tides. As the leaves of sea grass die, they add organic matter to the sea bottom that enriches it and provides food for the many detritus feeders. These miniature forests also provide shelter for a host of mollusks, crustaceans, and small fishes. Some of the invertebrates attach themselves to the leaf surfaces; some glide over the leaves as they graze on the algae film; and others prowl along the bottom among the plant stems or simply rest on the bottom.

Swimming Clams

Scallops are probably the most common and visible of the shellfish that rest among the *Zostera* plants. The several species of *Pecten, Aequipecten, Chlamys,* and others of the inshore scallops find beds of sea grass to be near-perfect habitats. Rhythmically pumping the nutrient-rich water between their two ribbed shells, scallops extract the food particles and oxygen they need. The shells, one flatter than the other, strewn over the sea bottom, resemble flower petals in hues of white, orange, purple, and brown, many accented with vivid, radiating orange bands. Their slightly parted edges reveal the scallop's 30 to

76–77. Scallops are bivalve mollusks, with shells composed of two parts hinged together at the back. Unlike other bivalves, scallops have many eyes, each complete with cornea, lens, and retina. Between the blue eyes of the Atlantic bay scallop (Aequipecten irradians) are numerous tentacles containing tactile and chemical sense organs. Scallops sometimes swim, though erratically, by rapidly opening and closing their valves.

There are three species of manatees (Trichechus): one along the coasts of the tropical western Atlantic; another in the Amazon and Orinoco drainages of South America; and a third in West Africa. Manatees are mammals that reach a large size—4 to 5 meters long and as much as 350 kilograms. Strict vegetarians, they eat virtually all aquatic plants, and even some terrestrial ones overhanging the water. In South America they have been used to clear canals of clogging vegetation. Very sensitive to changes in water temperature, manatees may die during very cold weather.

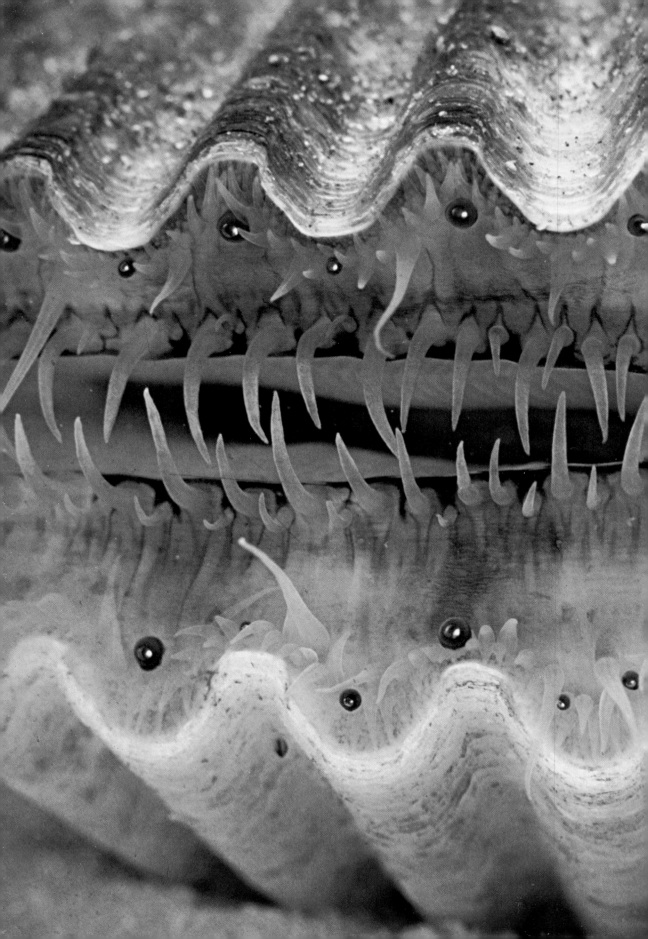

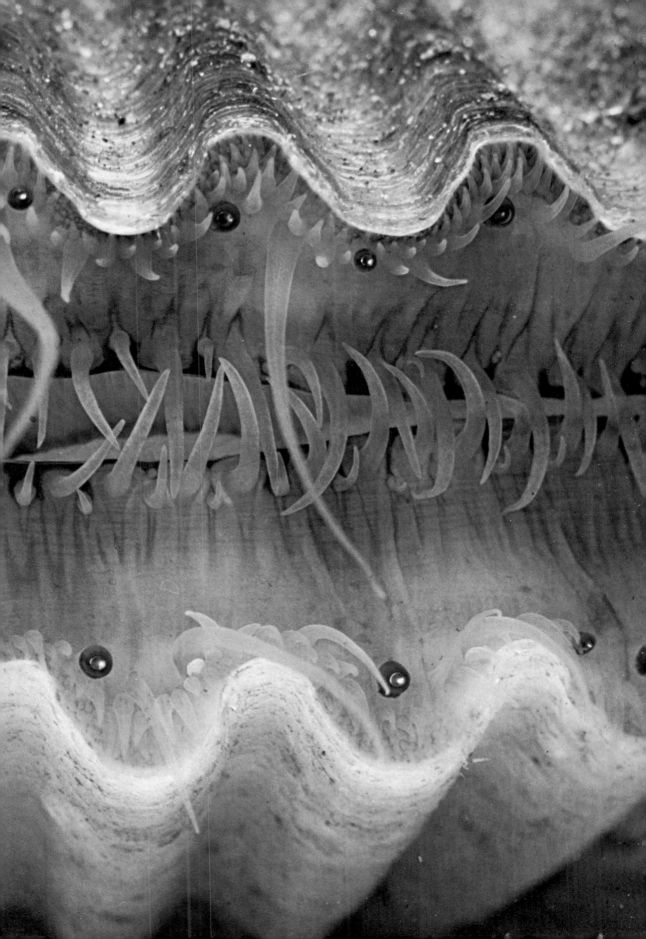

40 bright blue eyes, which seem to glow. Delicate feelers between the eyes are sensitive to any slight disturbances, especially the approach of that relentless predator, the starfish.

Scallops have been called swimming clams because they are able effectively to escape their enemies by "swimming" in a fluttering zigzag, like underwater butterflies. This distinctive movement is accomplished by rapidly opening and closing their shells, which shoots a small jet of water out between their edges. By this means, the animals may actually "swim" as much as a meter.

Clams also burrow in the soft bottom of the shallows zone. The burrowing clams are filter feeders and dig themselves in just below the surface of the sandy or muddy bottom, with paired siphons extending into the water above. Water is drawn in through one siphon, food particles are filtered out, and waste water is discharged, together with any other wastes, through the other siphon. Although they do not have the eyes or feelers of scallops, the burrowing clams are able to sense, and possibly taste, the presence of starfish.

The plain-looking *Macoma* clams are typical of this group and have a worldwide distribution. *Venus* or *Mercenaria* clamshells have a smooth inner surface that has a luminous, pearly luster. Rich purple and violet tones appear on the outer margin. One species of this group, *Mercenaria mercenaria*, common off the east coast of the United States, was used by the American Indians for wampum, a form of bead money made of polished shells. Purple wampum was the most valuable.

All of the *Venus* clams have been used as food by man at one time or another. Perhaps the most widely used and most valuable variety is the hard clam, or quahog (*M. mercenaria*), which brings more than $25 million a year to fishermen who harvest this tasty mollusk.

Pincushions of the Sea

Clams are common in areas where the water barely covers the bottom at low tide down to depths of 15 meters. In many places they are found in company with sea urchins, in a sort of underground habitat.

Anyone who has gone wading or skin diving in the shallows zones of tropical and subtropical oceans fears and respects the hatpin urchins (*Diadema antillarum*). These formidable-looking animals bear spines, as much as 30 centimeters long, that are black and resemble needles or long, old-fashioned ladies' hatpins. Marine biologists believe these hollow spines may be filled with poison.

If an unwary diver steps on a hatpin urchin, its spines easily penetrate his skin and break off. The victim quickly feels a burning pain around the embedded spine and, to his dismay, finds that trying to extract it is futile. Each spine is fitted with barbs, and the limy material composing it crumbles when grasped in a forceps.

The most formidable-looking—but actually harmless—sea urchin is the slate-pencil urchin (*Heterocentrotus mammillatus*) of the Indo-Pacific and Hawaiian areas. Its spines are slightly flattened, about 1 centimeter in diameter and 13 centimeters long. The interiors are very

Above. *Red and black West Indian sea urchins* (Diadema antillarum; Echinometra lucunter) *are exposed on a stony reef at low tide.*

79. *The slate-pencil sea urchin* (Heterocentrotus mammillatus), *occurring from the Red Sea to Hawaii, reaches a large size, with a shell (test) 15 centimeters in diameter and spines about the same length. It is found most often in association with colonies of living coral.*

Unlike sea stars, brittle stars and
basket stars can move rapidly
over the sea bottom by thrashing
their arms. The arms join at the
central disk, containing the
mouth and visceral organs, and in
some species branch many times.
When handled, the arms
sometimes break. Above. The
basket starfish (Gorgonocephalus
euenemis) is an intricately
fashioned creature that opens
only in the darkness. Right. The
underside of a brittle star
(Echinodermata ophiuroidea)
shows the animal's star-shaped
mouth. This species occurs in
Caribbean waters near Jamaica.

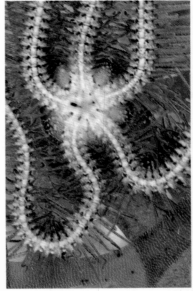

limy, hard, and white and can be used like chalk to write on slate blackboards. The spines also make interesting curios, and from them many island artisans fashion pendants, earrings, and other items of jewelry for sale to tourists.

Most sea urchins are vegetarians, browsing on algae. Some, however, are omnivorous predators that blindly forage about for shellfish, tube-building worms, crustaceans, even other echinoderms. Such food is quickly chewed with the efficient mouth apparatus called "Aristotle's lantern," which received this curious name because it was first described by the famed ancient Greek philosopher-naturalist as resembling the horn plates of a lantern.

Some sea urchins are used as a source of human food. Caribbean islanders collect the "sea egg" of *Tripneustes ventricosus* just before the animals spawn, when the orange ovaries are swollen with eggs. They crack open the shell of the urchin and eat the ovaries raw, roasted in its shell, or fried like a scrambled egg. In the Mediterranean region, Italians eat *Paracentrotus lividus*—in Italian, *frutta di mare*—with gusto.

A Multitude of Starfish

The shallows zone supports many kinds of starfish and their relatives, with nearly infinite differences in color, shape, and way of life. Some of the common starfish, *Asterias*, for example, are found from the shoreline down to depths of 50 meters. Frequently they make excursions into the littoral zone to prey on the attached shellfish. Others, such as the sunstars (*Solaster*) and the basket stars (*Gorgonocephalus*), which look nothing at all like the common five-armed starfish, do not enter the littoral zone. Sunstars may have as many as 24 arms and measure 60 centimeters in diameter. They prey on smaller starfish, sea anemones, clams, and snails.

Perhaps the handsomest sunstar is *Crossaster papposus*, found in the northern Pacific Ocean as far south as Vancouver Island and, in the Atlantic Ocean, down to New Jersey on the western side and the English Channel on the eastern. It has from 8 to 15 arms, each tufted with spines, and its upper surface displays a warm sunburst of colors—yellow, orange, pink, and red.

The grotesque-looking basket star resembles a woven reed basket that has begun to unravel. The animal has five arms, as does the common starfish, but each arm branches and rebranches until the entire organism looks like a confused mass of tentacles.

Gliding with sinuous movements of its fragile arms, the brittle star feeds on organic scraps, carrion, and other sea-bottom materials. The arms of some brittle stars glow in the dark. *Ophiacantha bidentata*, which inhabits both the Atlantic and Pacific, south to Portugal and South Carolina and down to California and Korea, appears dark brown in daylight but becomes a ghostly bluish-gray at night.

Crabs of Many Colors

The shallows zone is home for a variety of crustaceans: crabs, shrimps, prawns, and lobsters. The colorfully

82–83. *Two decorator crabs* (Macrocoeloma trispinosum), *from the tropical waters around the Florida Keys, carry different-colored sponges on their backs to serve as camouflage for confusing predators.*

Some sea stars have as many as 50 arms. Top. *The spiny sunstar* (Solaster papposus) *from northern waters has 15 arms.* Bottom. *The California sunflower star* (Pynopodia helianthoides) *may have as many as 24. Most sea stars are of moderate size—20 to 30 centimeters. Sunflower stars may grow to more than a half meter.*

arrayed blue crab (*Callinectes sapidus*) bears a shell as much as 20 centimeters across that is dark blue or blue-green to brownish with mottling of blue and cream. In the eastern Atlantic, the blue crab is found in the shallows of France, Holland, and Denmark; in the western Atlantic, it scuttles about on the ocean floor from Cape Cod to the Gulf of Mexico. The green crab (*Carcinides moenas*), which grows to about 7 or 8 centimeters across, is generally dark green or green with yellow mottling. A native of European waters, it was introduced into western Atlantic waters in the early 1800s.

These two crabs are eagerly sought by man, sometimes for different reasons. The blue crab has excellent flavor and is taken in pots by commercial fishermen and with long-handled nets by sportsmen. The green crab, a serious predator of young clams, oysters, and scallops, is trapped to reduce the damage it does to shellfish stocks in northwest Atlantic waters.

Omnipresent inhabitants of the shallows zone throughout the world ocean are the hermit crab (*Pagurus, Eupagurus*) and its relatives. This little crab looks as if it were only partially finished: the head, legs, and forward part of the body look like those of any other crab and are protected with a hard shell; the rear part, however, is soft and unprotected and can be quickly snipped off by any predator that chances on the hapless crab. To compensate, hermit crabs have a unique and quite successful way of protecting themselves; they put on a suit of armor over their vulnerable posterior—by seeking out an empty shell from a dead marine snail and neatly backing into it! In Atlantic waters the shell may be that of a whelk or winkle (*Busycon*), a dog whelk, or other snail. In the Indo-Pacific realm, the shell may be from a cone snail (*Conus*). Whichever it is, the hermit crab scuttles over the ocean floor dragging its shell home behind it, like a camping trailer. As the crab grows it must find a larger shell; like a person shopping for new clothes, it tries one shell after another until it finds one that fits and then settles comfortably into its new home.

In a curious natural alliance between hermit crabs and sea anemones, the crab deliberately attaches the soft-bodied, tentacled animals onto its shell home. When the crab changes shells, it moves its anemone tenants along with it. The anemone's stinging tentacles protect the crab from fish and other predators; the anemone, in turn, gathers scraps of food from the crab's messy feeding. Nobody knows for certain how the two kinds of animals first established their relationship, or even which of them made the first move. But it is a successful arrangement, and the odd relationship serves yet another purpose: with its tentacles, the stalked anemone helps camouflage the crab so that it resembles a rock. Such skillful deception is practiced by a number of crab species, some of them called "decorator crabs" because they attach bits of sponge or algae to their shells. In European waters, the sponge crab (*Dromia vulgaris*) often takes a bit of brightly colored sponge and drapes it over the top shell like a cloak, which it holds fast by means of a pair of highly modified legs. Crabs tend to be solitary animals, but an outstanding

84. *Crustaceans, the main food organisms of the ocean, have devised varied ingenious ways of avoiding predators. Hermit crabs, for instance, have modified the abdomen so they can fit into cast-off snail shells.* Top. *A hermit crab* (Paguristes maculatus), *temporarily without a shell, is housed in an orange sponge.* Bottom. *A blue crab* (Callinectes sapidus) *tries to escape by burrowing.*

Above. *A large hermit crab* (Eupagurus bernhardus) *prowls the sea floor, secure within the refuge he has established in a discarded whelk shell. Three commensal sea anemones* (Calliactus parasitica) *attached to the shell provide camouflage for the crab. The anemones feed on scraps of the crab's food.*

Above. *A large-clawed American lobster* (Homarus americanus) *prepares to defend its burrow by making threatening gestures with its powerful pincer claws. This species sometimes grows to 60 centimeters in length.*
Right. *The red little lobster* (Enoplometopus antillensis) *is a common decapod of many tropical waters.*

exception is the king crab (*Paralithodes camtschatica*) of the chill waters of the North Pacific and Bering Sea. When only about 3 to 5 centimeters in diameter, king crab juveniles climb upon one another until they form ball-shaped pods 3 to 4 meters across that may contain several thousand individuals. Scientists believe these red-orange, spiny little crabs may form the pods as a defense against sculpins and other predators. Later, when the crabs have grown to adults and are almost 1 meter from tip to tip of their outstretched legs, they move off into deeper water, safe from almost any predator.

Many Kinds of Lobsters

Unlike crabs, lobsters walk on the ocean floor in a straight-forward manner, not sideways. They scavenge over the bottom searching for dead fish, and because they have a very sensitive sense of smell, they can detect carrion from long distances. Should an enemy approach, they swim away, sculling through the water with powerful flexing strokes of the abdomen.

The most familiar-looking of the group are the homarid lobsters, the creatures with the large pincer-like claws. The European species is named *Homarus vulgaris*; the American form *Homarus americanus*. In the market, lobsters generally range from about 75 grams to perhaps 2 kilograms. Some very large American lobsters, weighing as much as 15 kilograms, have been caught in the shallows zone near the edge of the continental shelf. Homarid shells are mostly dark green, but some peculiar color combinations are also found, including mottled yellow and tan, blue, and half-blue half-green lobsters. However, no matter what their color when live, all of them turn an orangy red when cooked.

Homarid lobsters are found only in the North Atlantic. In the rest of the world ocean, the common lobsters are palinurids—the clawless, spiny lobsters. Palinurid lobsters lack the pincer claws but have a pair of extremely long and strong antennae, which they use as whips to discourage their enemies. Spiny lobsters may grow to weigh as much as homarid lobsters, but because they have been overfished, such patriarchs are seldom found nowadays.

Most lobsters avoid their fellows because they are cannibals, and a chance meeting of two lobsters on the bottom often means that one ends up as a meal for the other. But the spiny lobster of the Caribbean (*Panulirus argus*) is different; it has attracted the attention of marine biologists because it carries out a peculiar "queuing" behavior. In the autumn, when the shallows zone chills, the lobsters form long lines, with each animal touching the abdomen of the one in front of it, and march off together into deeper water. Scientists are not certain why the lobsters form these queues, nor what benefit the animals may derive from such a curious aggregation. But their urge to follow head-to-tail is very strong. Underwater observers have seen the long lines of lobsters solemnly marching in single file over the sandy bottom, past rocky outcrops where other lobsters have already taken refuge. Without hesitation, as if a Pied Piper were beckoning,

88–89. *The common octopus* (Octopus vulgaris), *usually about 30 centimeters arm tip to arm tip, is a fearsome-looking creature. The octopus, though often the subject of horror stories and sometimes called "devil fish," is a shy, retiring animal that generally feeds on fish and crabs.*

91. *The Atlantic cuttlefish (Sepia officinalis) has 10 "arms," or "feet" that surround the head. Within the body there is a hard, limy, flat shell that serves as a skeleton. The internal shell contains gas-filled spaces; like the nautilus, cuttlefishes regulate their buoyancy by changing the amount of gas in the shell. Light causes a decrease in buoyancy, and the cuttlefishes burrow into the bottom during the day. At night they are active swimmers, but not nearly so strong and fast as the squids.*

Below. *One of the most interesting invertebrates is the chambered nautilus (Nautilus pompilius). Resembling many different extinct (fossil) cephalopods, it is, in effect, a living fossil. Unlike many other shelled marine creatures, the nautilus is a strong swimmer. Its great buoyancy is attributed to gas secreted into the empty chambers of its shell.*

the lobsters in the caves quickly scramble out of their refuges and join their fellows in the long, single-file parade.

A Deadly Embrace

Everywhere in the Atlantic Ocean that crabs and lobsters are prevalent, one of their chief predators, the common octopus (*Octopus vulgaris*) is prevalent, too. Homarid lobsters can defend themselves against the octopus by using the crushing power of their large, well-muscled pincer claws. The spiny lobster, however, is almost helpless before the persistent octopus, which can ooze into the most secure crevice a lobster chooses for shelter. The octopus virtually smothers the lobster in its fatal eight-armed embrace and proceeds to tear it apart with its sharp parrot-like beak.

Cephalopods—the squids and octopuses—are in fact among the most active hunters in the shallows zones of the world ocean. Darting forward and backward in the water like squadrons of sleek jet aircraft, the squids (*Illex, Loligo*) snatch at drifting food scraps dropped by larger predators. When small mackerel and other juvenile fishes approach, the squids become very active and very proficient predators. They stalk their prey and then, in a final short rush, wrap their arms about it in a deadly embrace and start to feed. While *Illex* and *Loligo* may grow to about 90 centimeters long, other squids range in size from one-half centimeter (*Sandalops*) to 18 meters (*Architeuthis*).

In the warmer waters of the world, the active cephalopods are the cuttlefishes, which are most abundant in the Mediterranean, the western Pacific, and the Indian Ocean. The best-known of cuttlefishes is *Sepia officinalis*, which preys in Atlantic waters, over the shallows zone from the English Channel to the Cape of Good Hope. Fishermen in the North Sea sometimes bring up this multi-armed animal in their nets; in the Mediterranean it is trapped in small earthen pots, *mummarellas*, that fishermen set out on the sea bottom.

It is difficult to think of cephalopods as attractive. In fact, the octopus has figured in so many horror tales of the sea—and wrongfully portrayed, at that—that the very sight or mention of this retiring animal causes most people to shudder. But two groups of cephalopods are considered by some shell collectors as jewels of the sea. One group includes the argonauts, especially the paper nautilus (*Argonauta argo*), found in the surface waters of all warm seas of the world. This graceful animal is about 30 centimeters long and has a delicate, pearly white, fluted spiral shell. The other group includes the chambered or pearly nautilus (*Nautilus pompilius*), common in the Indo-Pacific realm and resembling the paper nautilus (though the two are not related). The spiral shell of the pearly nautilus, instead, is a strong-walled floating home of a creamy hue with distinctive cocoa-brown stripes.

Flying Fish Traps

Cephalopods and the crustaceans and small fishes on which they feed all too often become victims of the many birds

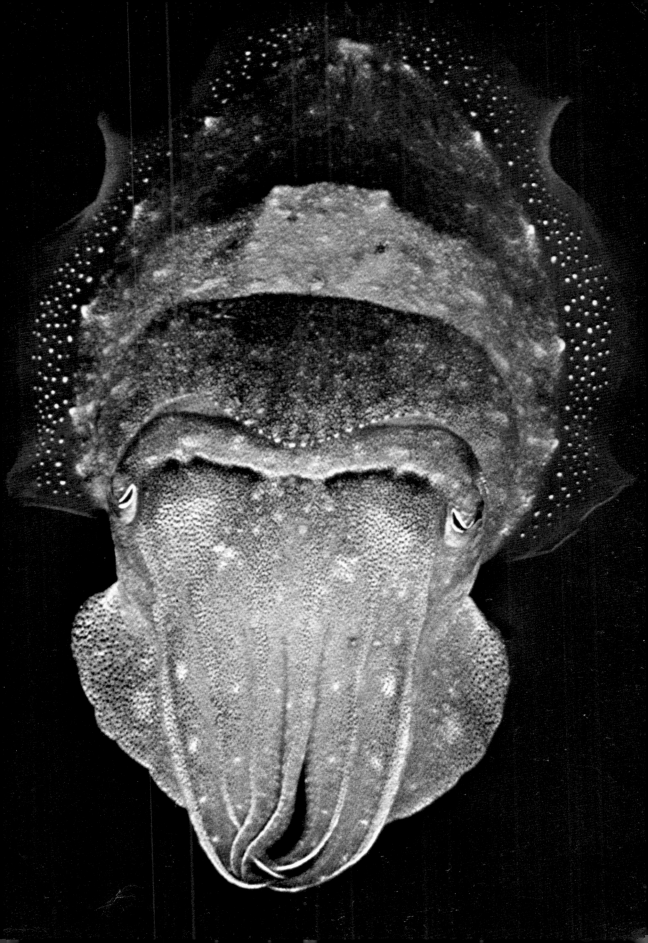

The brown pelican (Pelecanus occidentalis) seemed headed for extinction in North America during the late 1960s and early 1970s. During this period, many pelican eggs had abnormally thin shells that broke before hatching, and many others contained dead embryos. High levels of DDT-type pesticides were found in these shells and embryos. Contamination of the food chain from agricultural use of pesticides appears to have been the cause. Since the ban on DDT, there has been a slow but significant decrease in the pesticide content of eggs and an increase in successful reproduction of pelicans. Once again, family groups can be seen hunting together along the shores of the Gulf of Mexico and the Gulf of California, where the birds breed on islands. Brown pelicans locate their prey by sight and, thus, are usually found in areas of clear water. When potential prey such as a school of jumping fish is sighted, the birds dive from a height of about 10 meters above the water. In this diving behavior and in their strictly coastal habitat, brown pelicans differ from their nondiving and lake-dwelling relatives, the white pelicans.

that soar and glide over the shallows zone in an unending search for food. From their advantageous aerial viewpoint, the birds can prey and swoop down upon the unsuspecting marine dwellers with deadly accuracy.

In tropical and subtropical waters, when the brown pelican (*Pelecanus occidentalis*) sights a fish, it makes spectacular headlong dives from heights of 3 to 5 meters, entering the water with a resounding "plop" that can often be heard as much as a kilometer away. Pelicans often scoop up small fish that have been driven in toward shore by sharks and hold them in their flabby-looking throat pouches.

Many a yachtsman or fisherman has marveled at the precision dive-bombing assaults of the gannets (*Morus*). These large white seabirds, with their distinctive yellow bill, hover on beating wings as much as 30 meters above the restless surface of the sea, ceaselessly watching for fish in the depths. At the first sign of prey, the gannets fold their wings and hurtle into the depths, breaking the surface with barely a splash. The momentum of their dive carries them nearly 10 meters undersea, and with feet and wings in motion, the birds can "fly" underwater after their targets.

Many seabirds are best able to capture their prey by swimming skillfully through the depths. Loons and other divers swim to depths of 50 to 70 meters in pursuit of prey, often remaining submerged for 15 minutes. The black-throated diver (*Gavia arctica*), like many human tourists, spends its winters in the warmth of the Black Sea and summers in the Baltic Sea and in northern Siberia. Its relative, the common loon (*G. immer*), takes part in the same sort of seasonal migration between the Gulf of California in the west, the Gulf of Mexico in the east, and the subarctic portions of North America.

Joining the onslaught against the fish and crustaceans of the shallows zone are the many small but active birds that tenant the rocky crags and ledges at the edge of the sea. The Atlantic puffin (*Fratercula arctica*) is one of the most striking of these. Only 25 centimeters tall, it has a large, well-sculptured red bill that sits on the front of its head like the red comb of a rooster. The puffin is an efficient fisher and returns to its cliff perch with a small fish held firmly in its bill. The dovekies (*Plautus alle*), the murres (*Uria*), and the auks, particularly the razorbilled auk (*Alca torda*), all active exploiters of small fishes, skillfully hunt them out and pursue them as they dart and twist through the water in an effort to escape.

A Wealth of Fishes

In springtime, when the plankton "bloom," the herring, mackerel, and other feeders begin to move over the shallows zone from their wintering grounds in the open sea. The darting shoals of herring move through the waters, each fish snapping at individual *Calanus* animals. Like children turned loose in a sweet shop, they snatch at their prey as quickly as they can. Soon the herring grow fat from their voracious feeding and fishermen spread long curtains of net in the sea to trap the plentiful schooling fish.

Other predators quickly take advantage of the silvery abundance. In the eastern Atlantic, cod (*Gadus morhua*), urged on by the need to spawn, migrate from subarctic waters into the North Sea. Moving through the chill, black depths, they approach the coasts and in March begin to spawn. The fertilized eggs drift upward through the water to the surface layers, where they will develop. Quickly the herring turn to the tiny bobbing spheres and snatch at them until it seems almost as if not a single cod egg will survive. But nature has endowed the cod with a vast supply of eggs—9 million to 15 million per female—and enough escape the rapacious herring and other predators to replenish the species.

Many of the herring, replete with cod eggs and, later, the developing fry, swim swiftly down into the gloomy depths. There they encounter the cod, eager to restore their own strength after the fast of spawning. Now the cod feast on the well-fed herring; as they glut themselves on the darting hordes, they pause occasionally to pick clumps of developing herring eggs from the bottom. The cod also are threatened by man, the ultimate predator. Baited hooks laid on the bottom and gill nets anchored to the sea floor take their toll of the cod. Immense trawl nets, stretching 20 to 25 meters at the mouth and 35 meters long, engulf everything in their path, catching hundreds of cod, herring, gurnards, and skates.

But at the wave-tossed surface, the cod eggs continue to develop and hatch into popeyed larvae that soon prey on minute animals. Larvae of barnacles, lobsters, shrimps, and crabs, as well as copepods, fall victim to the active little predators. Later, when the young cod are about a month old and perhaps 20 to 25 millimeters long, many form a relationship with the lion's-mane jellyfish (*Cyanea*). The floating bell of this colonial sea animal may measure nearly 2 meters in diameter, but is usually 30 to 60 centimeters. Long, deadly, stinging tentacles hang down from the bell, ready to trap and kill the small creatures the jellyfish depends on for food. Several tiny cod can often be found swimming with apparent immunity among the tentacles. Scientists have found that the codling and jellyfish exist in a mutually beneficial arrangement. The little fish rid the *Cyanea* of a burrowing parasite and are in turn protected from whiting (*Gadus merlangius*) *Merluccius bilinearis*) and other enemies. But life under the jellyfish is not some codfish utopia. When occasionally the codlings swim too close to its bell and are killed, the *Cyanea* loses no time in making a meal of its former partner.

Later in the season, as summer warms the topmost part of the sea, the young cod hatched from the floating eggs swim down into the chill blackness of the depths, to seek out the small animals that will nourish their growth. But there also they are stalked by a host of predators, including their own kind. The dogfishes, *Scyliorhinus* and *Squalus*, particularly the latter, join in the feeding orgy and attack herring and cod alike. The diminutive dogfishes, usually 1 meter or less in length, mirror in miniature the voraciousness of their larger brethren in the fearsome shark group.

96–97. *A sand tiger shark* (Odontaspis taurus) *opens its mouth to reveal the deadly teeth that put the animal at the top of the food web. This species ferociously attacks and consumes fish, including other sharks, and marine mammals. It is considered dangerous to swimmers, although there are no reliable records of attacks on man.*

94. *Mexican scad* (Decapterus hypodus) *form dense schools. The reason for the schools remains a mystery; perhaps they serve as defense against predators. Scads are related to the jacks and pompanos.*

Above. *Tufted puffins* (Lunda cirrhata) *are fish eaters that breed on rocky coastal islands as far south as the Farallon Islands near San Francisco. Using the stout claws on their webbed feet, they scratch out a burrow in a rocky crevice, where the female lays a single large egg. The chick remains in the burrow, protected from gulls and jaegers until it is ready to fledge.*

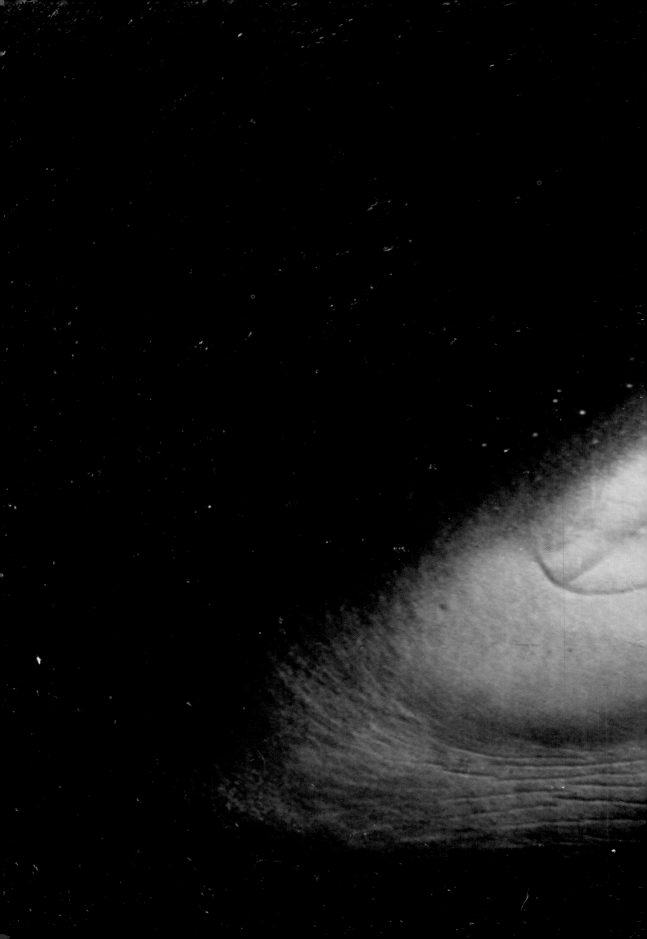

Stalking the ice floes and frigid waters of the Arctic in search of seals, the polar bear (Ursus maritimus) is one of the great nomads. Top. A female and her cubs swim through the chill waters. Bottom. A large polar bear, nearly 2½ meters long, runs easily over the pack ice. These massive beasts can run at a speed of 65 kilometers per hour.

Feeding Machines: The Sharks

The productivity of the shallows zone and the abundance of fishes and other marine organisms there are most effective lures for some of the most active and efficient predators in the world ocean. The most feared and destructive are the large sharks, whose sleek bodies torpedo through the water and explode the packed shoals of herring and mackerel. The porbeagle (*Lamna nasus*) drives its 3-meter-long body with powerful sweeping thrusts of its tail fin to slash again and again at the tightly milling schools. So eager is the porbeagle that it will attack fish already meshed in fishermen's nets, tearing the netting and ripping the fish to bloodied shreds. Other sharks may join the onslaught, including the bizarre hammerhead (*Sphyrna*), with its eyes and nostrils at opposite ends of the grotesquely shaped head that gives it its name. Other sharks feeding on the abundance of the area present a curious paradox of form and function. One is the basking shark (*Cetorhinus maximus*), whose chocolate-brown body may measure nearly 13 meters long. Another is the whale shark (*Rhincodon typus*), the largest fish in the world, whose spotted body may reach an awesome length of 15 to 16 meters. The sight of these great beasts swimming may strike terror in the hearts of fishermen and boaters, but their frightening appearance belies their actual habits. Both of these gargantuan fishes are gentle giants that pay little if any attention to men and nearby boats. Occasional sightings at sea of these giant sharks and, especially of their stranded, partially decomposed carcasses or bleached skeletons no doubt have contributed to numerous reports of "sea serpents." Another predator in shallow waters is the blue shark (*Prionace glauca*). Almost indigo on its upper surface, this shark becomes a gleaming white toward the underside. It has a sharp nose, a slender body, and a long upper tail lobe. Most blue sharks are no more than 3 meters long. They devour a great many small fishes, including their own kind.

Hunter and Hunted

Because of the abundance of organisms found there, the shallows zones of the world ocean are very active arenas, where predator and prey carry out their roles of hunter and hunted, to kill or be killed. At the top of the list of predators are the seemingly invulnerable marine mammals. Yet, they too cannot escape destruction. The hooded seal of the North Atlantic (*Cystophora cristata*) and the northern fur seal of the North Pacific (*Callorhinus ursinus*) grow sleek and fat on a diet of squid and fish, especially herring. But the hooded seal may fall victim to the fearsome attack of the polar bear (*Ursus maritimus*), which is as much at home swimming in chill arctic seas as it is plodding over frozen polar barrens.

The great white polar bear, king of the icebound Arctic Ocean, stalks majestically over the floes or swims across open water in search of its prey. Eskimo lore says an actively hunting polar bear can smell a seal 35 kilometers away. This bear's hunting success is due to both its natural camouflage and its stalking techniques. Its white coat blends well with the white of the arctic ice and snow. And when a bear creeps up on a seal basking on the ice, it is said to push a chunk of ice ahead of it with its nose for concealment. When within striking distance of its prey, the bear pounces or lashes out with its powerful paws and can easily kill a 200-kilogram bearded seal. Or, a polar bear may lie in wait beside a seal's breathing hole, ready to pounce on the surfacing seal and retrieve it through the smashed ice. Using yet another method, the bear may ease itself into the water with scarcely a ripple and swim underwater or with only the tip of its black nose showing, later to surface beside a sleeping seal.

The polar bear's size helps it to handle large seals with as much ease as a cat catches mice. A mature male may weigh half a ton and measure 2½ meters in length and 1½ meters high at the shoulder. Standing erect for a wider view of its kingdom, a polar bear can rear up to 3½ meters. These massive beasts can run 65 kilometers per hour, swim 10 kilometers per hour, and travel through the water nonstop for 200 kilometers.

In its yearly journeys of a thousand kilometers from the wintering grounds of California to the Pribilof Islands of Alaska, the Alaska fur seal runs a gauntlet of enemies. On the Pribilofs, where the seals haul out to bear their young and mate, the massive bull seals, some weighing 1,200 kilograms, fight to establish breeding territories and to gather harems. A few bulls die of injuries and the smaller females often are injured or killed in the raging battles. Mortality among seal pups from injury, accident, and disease is high, too.

Predators take their toll as well. In the relatively warm waters that flow past California, great white sharks patrol in a perpetual search for prey. A female fur seal weighing only about 200 kilograms is no match for a hungry shark able to bite the seal in two with very little effort. But the most fearsome predator of all is the killer whale, which actively pursues the fur seals nearly throughout their range and which may swallow them whole.

An attack on a small herd of swimming fur seals by a pack of five killer whales was graphically described by Victor B. Scheffer, an American marine mammal expert. He tells how a large bull killer whale swam beneath the surface like a torpedo for 5 minutes, to suddenly reappear with a young male fur seal held sideways in its jaws. Like a cat with a mouse, the killer whale played with the fur seal for 20 minutes, tossing it into the air, catching it, and thrashing it about. Finally, perhaps tiring of its play or else feeling pangs of hunger, the killer whale bit deeply into the helpless seal's body, and then swallowed it whole.

Seals, and sometimes polar bears, may be attacked by the walrus (*Odobenus*), not necessarily as prey but as interlopers in the great tusked beast's territory. This territory includes the ocean floor, in the chill depths 60 to 90 meters below the surface. Here the walrus, which may weigh 700 to 1,400 kilograms, efficiently plows the bottom with its ivory tusks to uncover the bivalve mollusks and worms that are its food. Occasionally, a walrus may

100–101. Pinnipeds must come ashore to breed. Breeding colonies tend to occur on isolated islands in the higher latitudes. The male of the species, larger than the female, maintains a harem. Seals can remain underwater for as long as 20 minutes and can dive as deep as 300 meters. Row 1: left, Galapagos fur seal (Arctocephalus galapagoensis); center, Weddell seal (Leptonychotes weddelli); right, crabeater seal (Lobodon carcinophaga). Row 2: left and center, harp seal (Pagophilus groenlandicus); right, Steller's sea lion (Eumetopia jubata). Row 3: left and center, harp seal; right, Steller's sea lion.

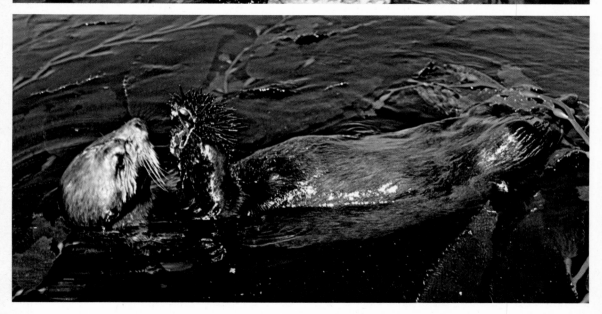

102. *The sea otter (Enhydra lutris) was once hunted to near-extinction for its luxurious fur. Today, as a result of conservation practices, this intelligent marine mammal has increased in abundance in the Pacific Ocean, off the coast of the United States.* Top. *A sea otter removes a sea urchin (Echinoidea) from a stipe of kelp* (Macrocystis). Center. *Lying on its back at the surface, the sea otter smashes a clam (Pelecypoda) against a rock held like an anvil on its stomach.* Bottom. *Resting at the surface among kelp fronds, the sea otter dines on a sea urchin.*

104–105. *The walrus (Odobenus rosmarus) is a pinniped, classed in the family Odobenidae. The species inhabits the Arctic Ocean and migrates with the advance and retreat of the ice. Walruses are strong swimmers but ride on ice floes during migration. They forage on the ocean bottom at depths to 90 meters. Their upper canine teeth grow throughout life and form tusks, which may reach 100 centimeters in males and 60 centimeters in females. These tusks are apparently used to probe the bottom for shelled mollusks, which constitute their staple food. How the walrus extracts the mollusk from its shell remains a mystery, since no crushed shells have ever been found in a walrus stomach.*

devour a seal it has attacked. Tearing the seal to pieces with its bristle-covered mouth pads, the walrus swallows the chunks of flesh.

Killer whales sometimes pursue walrus among the ice floes of arctic waters and the Bering Sea, but the greatest enemy of this distinctive tusked beast is man. For thousands of years, Eskimos hunted the walrus as an absolute necessity for survival. Its flesh supplied them with food, its ivory tusks were made into spear and harpoon points, its skins provided covering for tents and boats as well as leather thongs for dogsled harnesses and fishing lines, its intestines were slit and sewn together for rain garments, its stomach contents were eaten, and its teeth and tusks were traded or carved as curios.

Return of the Sea Otters

The history of man's exploitation of marine mammals is a long and sordid one. Some animals have been hunted to extinction. Others, such as the walrus, are now considered threatened species. A few have been brought back to something like their former numbers. One such animal is the sea otter.

Sea otters are unique because they use tools in their feeding. Their food-gathering takes place on the ocean floor, where they forage as deep as 50 meters for crabs, sea urchins, bottom fish, and octopus. Grasping the food in monkey-like paws as they float on their backs at the surface, they crush and tear the prey with powerful teeth. But occasionally an otter will place a stone or heavy shell on its chest and smash shellfish against this "tool" in order to get at the succulent meat inside.

For a long time, observers puzzled over why sea otters frequently roll over at the surface. But careful examination of the animal, especially its fur, revealed the reason for the rolling. Unlike most other marine mammals, sea otters do not have a thick layer of insulating fat under their skins. In order to retain body heat and stay warm in the frigid waters that are their home, they roll to trap air among the long hairs of their fur. These air bubbles serve as an insulating blanket.

The Danish explorer Vitus Bering, then working for the Russians, discovered both the Aleutian Islands of Alaska and the sea otter in 1741. At the time, sea otters ranged around the rim of the North Pacific from Japan along the coast of North America to Baja California. The thick, rich brown fur of these 20-kilogram animals proved their undoing. It became valued ornamentation for winter clothing, and the prized animals were virtually wiped out by hunters. Fortunately, enlightened modern conservation efforts have restored the sea otter to much of its former range. Today, boaters and visitors to the kelp beds over the shallows of the North Pacific can watch these playful creatures, with their sad-looking, old-men's faces, lolling in the gentle ocean swells.

Coral Reefs

The massive Great Barrier Reef of Australia, the delicate atoll reefs of Bikini Island and other South Pacific dots of land, and the network of fringing reefs of the warm waters of the world ocean support a varied, complex and colorful assemblage of marine animals. The corals forming the reefs are actually the skeletons of living animals that resemble miniature sea anemones.

Corals can be found at all depths in the ocean and in most geographical areas, including the chill North Atlantic and even the gloomy depths of Norwegian fjords. But reef-building corals are found only in those shallow tropical seas that are well lighted by the sun and where the water temperature rarely falls below about 20°C. Thus, coral reefs are widely distributed in the tropical Pacific Ocean, but in the Atlantic Ocean they are confined to the offing of the Florida Keys, around the Bahamas and in the island-dotted Caribbean Sea.

Coral reefs provide both food and shelter to the myriad brightly colored fishes that swim through the nooks and crannies of the reef structure. The colorful parrot fish munches with massive jaws and chisel-like teeth on coral, digesting the soft polyps and expelling the lining material. Fierce-looking moray eels peer out of crevices, waiting for their prey to swim by. Throughout the reef, voracious sharks relentlessly patrol the depths, ever alert for a fish that is in trouble and is thus likely prey.

The reef is formed, among other organisms, of the coral animal, the polyp. Like its relatives the sea anemones and jellyfish, the polyp belongs to the zoological phylum called Coelenterata, derived from a Greek word meaning "hollow bowels." The polyp secretes lime and builds about itself a mini-fortress, a kind of stony skeleton that it cements to its neighbors. As the polyps multiply, their skeletons fuse into the massive structures we see, which may be branched, cup-shaped, or boulder-like. Some grow horizontally into lobes and flower-like forms; others grow by coiling and winding sinuously, as brain corals do. When the polyps grow in looser patterns, they form the graceful branching structures of the elkhorn and stag-horn corals.

Growth of the Coral Reef

Very few kinds of corals are solitary. Most are colonial animals, with hundreds to millions of individuals forming the massive reefs. A reef is started when drifting larvae that have fortunately escaped the multitude of plankton feeders settle down on a piece of hard substrate—on a rock, a long-dead coral head, or, as frequently seen in the Pacific Ocean battle zones of World War II, a sunken cruiser or tanker. When the larva attaches, it transforms into a hard polyp with a mouth and other organs and begins to secrete the skeleton. The depth at which reef corals grow is controlled largely by the transparency of the water—that is, by how much sunlight penetrates. However, some species of non-reef-forming corals have been found growing almost 6,000 meters below the water surface—well into the abyss.

The importance of water transparency and sunlight penetration in coral reef formation was a mystery until careful study of the polyps revealed a close association between the coral animals and microscopic algae cells, called *zooxanthellae*, within their tissues. It is believed the polyp is able to use the oxygen produced by the zooxanthellae.

Another group of plants helps the coral polyps convert the dissolved minerals in seawater into a stony reef fortress. Two groups of calcareous algae, or "seaweeds," incorporate calcium into their tissues, so that they feel stiff and crumble when touched. The limy material from the calcareous algae helps to cement and solidify the reef into a virtually impregnable fortress.

One such alga is *Halimeda*, which has leaf-like growths only a few centimeters in diameter. The other is *Penicillus*, which grows about 10 centimeters tall in sandy patches among the reefs and resembles an old-fashioned shaving brush, complete with handle and bristles. Indeed, *Penicillus*, is frequently called "Neptune's shaving brush" and seems fit for the sea god to lather his face with.

A Longer Day

Many coral formations resemble leafless trees in an underwater forest, and at one time it was thought the coral formations *were* the lime-encrusted trunks and branches of trees petrified by a flooding sea. Some researchers, intrigued by this fancied resemblance to trees, cut the coral "trunks" to see whether growth rings could be counted, just as they are in the stumps of oaks or pines. Though the researchers found no rings in the cross section, they did find rings on the surface of the coral stems, which resembled, in miniature, a stack of pancakes.

In 1963, J. W. Wells, a scientist interested in fossil corals, examined some specimens that lived in Devonian times, 370 million years ago. He found broad bands composed of numerous very narrow bands. When he examined the coral under a microscope, he found there were 400 narrow bands within each broad band. Studies of modern coral indicate that the broad bands (composed of 365 narrow bands) represent one year's growth. Thus Wells speculated that the Devonian year was 400 days long, each day being 22 hours long. Many scientists believe that because of ocean tides and their frictional interference with the Earth's rotation, the spinning of the Earth on its axis has slowed over the past 370 million years.

Coral Relatives

Not all the attached animals on the reef are true corals. A living reef is a complex mosaic of many other sessile organisms and coral relatives, including the soft corals, sea fans, deadmen's fingers, sea whips, and gorgonians. As we descend into the depths, below about 50 meters, few of the reef builders exist there, and the structure is dominated by sea whips and gorgonians, some as much as 2 meters tall. The gold leaf gorgonian (*Trichinogorgia faulkneri*) looks like a small tree branch with many delicate gold-plated twigs. It is a rare species

108–109. *Orange fairy sea basslets* (Anthias squamipinnis) *swim above a protective Red Sea coral head, which features a variety of colorful soft corals* (Alcyonaria) *and hard corals* (Madreporaria).

Corals are animals that live singly or in colonies. Each is a tube-shaped polyp with a closed foot at one end and a crown of tentacles surrounding the mouth at the other end. By extending its tentacles into the water, the polyp feeds on plankton. The tentacles are armed with stinging cells called nematocysts. Most polyps are small (1 to 3 millimeters). Among the stony corals that make up reefs, each polyp can retreat into its own little cup in the rock-like skeleton secreted by the colony. This solitary coral, with its extended polyp, is from Florida's west coast.

111. *On a reef in the Red Sea, both corals and basslets (Anthias) seem to bask in the sunlight from above. Reef-building corals need sunlight in order to grow. In the tissue of each coral polyp there are microscopic green plants called zooxanthellae. The relation between the plant and the polyp is an example of symbiosis—a living arrangement advantageous to both organisms. The polyp provides the plant with a place to live. The plant provides the polyp with food. Under conditions favorable to growth, the zooxanthellae multiply rapidly and overpopulate the polyp. The polyp "prunes" its garden by releasing excess zooxanthellae into the water.*

112–113. *The subtle, sometimes brilliant colorings of living corals make the coral reef as beautiful as a flower garden.* Row 1: left, *tube corals (Dendrophyllidae);* center, Turbinaria; right, *brain coral* (Diploria labyrinthiformis). Row 2: left, *bubble coral (Eusmiliidae);* center, *gorgorian* (Ellisella); right, *Scleractinia.* Row 3: left, *maze coral* (Pectinia); center, *pillar coral* (Dendrogyra cylindrus); right, *knob coral (Poritidae).*

and, like the true corals, is not a plant but an aggregation of animals, the polyps, housed in their communal skeleton. Danger as well as beauty can be found among the coral relatives. Many a diver collecting coral has found to his discomfort that he has handled or accidentally brushed against *Millepora,* the stinging coral. It has long slender polyps that possess powerful nematocysts which can penetrate human skin. The poison discharged by the nematocysts causes a burning sensation; thus *Millepora* is also called "fire coral." The careless diver will not quickly forget his brush with it. The swelling and itching caused by its poison often lasts for more than a week.

Types of Reefs

Scientists have classified coral reefs in three main categories. One type is the fringing reef, which, as its name implies, forms a narrow band parallel to the coast and, in places may actually be connected to the land. Most fringing reefs are from tens to hundreds of meters wide. Several of the Hawaiian Islands are surrounded by fringing reefs, which lure skindivers eager to explore the fantastic underwater gardens.

Another type of coral formation is the barrier reef, a massive and formidable rampart that often is separated from the landmass by tranquil lagoons. Here and there, the barrier reef may be pierced by passages through which skilled navigators can pilot their ships. The Great Barrier Reef of Australia is the world's largest and most spectacular example of the type. It is some 150 kilometers wide and extends for about 2,500 kilometers.

The third and most romantic-looking type of reef is the atoll, which embodies everything one imagines when the words "South Sea Island" are mentioned. An atoll is simply a lagoon about 50 meters deep surrounded by coral reefs. The outer portion of the reef may have developed low, carbonate sand islands topped with graceful coconut palms, a few shrubs and other low trees, and small populations of tropical birds and insects. The land plants and animals colonize the atoll in a variety of ways. Storm winds often blow birds—particularly far-ranging species such as the albatrosses—far out to sea. They alight on an atoll to rest and may remain there to nest. Some land birds are also blown out to the atolls during storms. Frequently these birds carry bits of tree and fruit seeds in their crops and spread these with their droppings, thus planting the atoll with a variety of vegetation. Some birds, insects, and a few small mammals may arrive on the atoll as chance passengers on a flotsam island broken loose from another atoll hundreds of kilometers away.

The most persistent colonizers are the coconut palms. The fruit of the tree has a thick, fibrous husk and can float for months without being soaked with salt water. When a floating coconut washes up on the beach of an atoll, it eventually sprouts and grows into a tall, graceful plume-topped palm. The endurance of the coconut is well illustrated by the record of one, complete with husk, that washed up on a rocky beach on the coast of Norway.

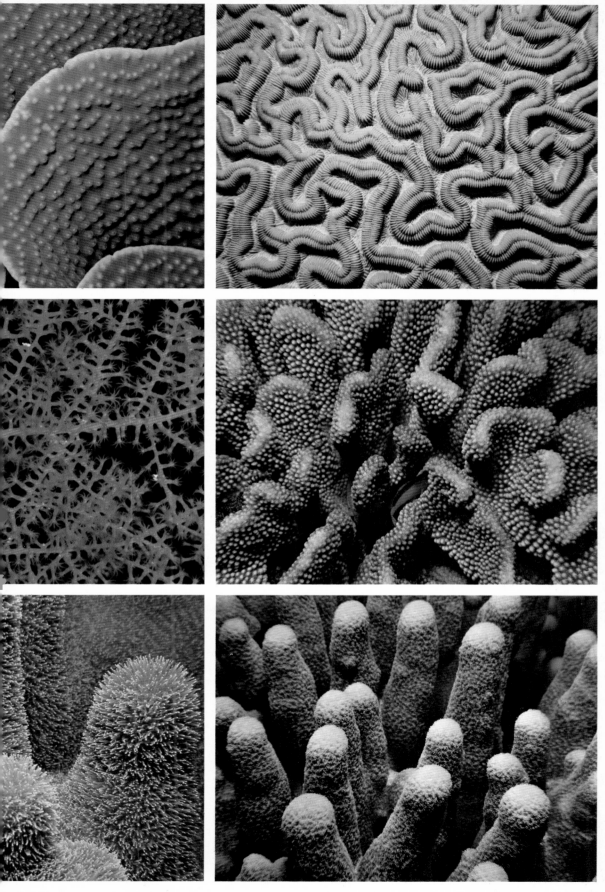

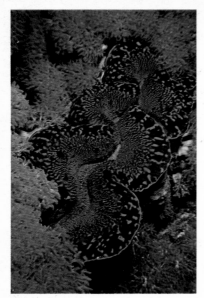

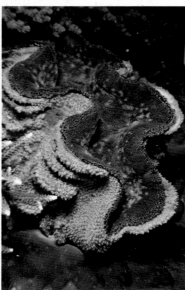

Giant clams (Tridacna) *are common in the reefs of the Red Sea,* top, *and the Palau Islands,* bottom, *as well as other islands in the Indian and Pacific oceans. Like corals, the clams have in their brilliantly colored tissues microscopic plants they expose to the sunlight, so that the plants can manufacture food.*

Presumably the drifting nut had traveled from an island in the Caribbean Sea, across the turbulent Atlantic by way of the Gulf Stream and the North Atlantic Current. A man picked up the nut and, his curiosity piqued, planted it in soil in a greenhouse. To his amazement, the coconut sprouted and grew.

Bizarre Reef Animals

A coral reef is a maze of openings that provide an ideal habitat for fishes, worms, urchins, and a host of organisms, all of strange hues and bizarre shapes. Partially buried among the coral formations in the tropical Pacific and Indian oceans, the giant clam (*Tridacna*) systematically pumps water in and out of its body, straining out the plankton on which it feeds. These gargantuan relatives of the familiar soft clam and scallop of temperate waters often grow shells that are 1½ meters across. The lips of the shell are formed with rounded ridges that fit together when the shell is closed, like the teeth of meshing gears. The soft part of the clam, the mantle, which extends past the lips of the shell when it is open, appears blue-green because of the dense colonies of zooxanthellae within the mantle tissues. Unverified tales are frequently told of pearl divers accidentally thrusting an arm or leg into the slightly parted shells of *Tridacna* and being trapped and drowned when the animal clamped its shell shut.

A skin diver's swim through a coral reef resembles a trip through some huge tropical fish aquarium. Schools of bright blue damsel fish (*Dascyllus reticulatus*) hover over the coral heads in an endless search for food. The orange-and-white anemone fish (*Amphiprion percula*) plays hide-and-seek among the stinging tentacles of its living refuge, while the multicolored angelfish (*Pomacanthus*) pokes among the coral branches for tidbits. The resemblance between a reef and a tropical fish aquarium should not be too surprising, since it is from such reefs that collectors capture many of the specimens sold to aquariums

For many years, scientists wondered about the reasons for the strange shapes and brilliant colors of the reef fishes. The reds, yellows, greens, and blues seemed to be almost a quirk of nature, since all fishes are believed to be color-blind, seeing their watery world only in shades of gray. What difference would it make to a barracuda or grouper whether the blenny it was attacking was blue or brown? The answer is that the color makes no difference. What matters is the way the prey fish appears to the attacker against the background of the reef. That is, if the shade of gray the blenny *appears* to be to the grouper blends with the apparent gray of the reef, the grouper is less likely to see the blenny. Thus camouflaged, the blenny is able to survive longer in the watery jungle. The markings, shadings, and stripes on a reef fish also add to its camouflage and help it blend into its background by "dazzling" the predator fish.

Predators on the Prowl

The popular image of predators on the coral reef is that of sleek-bodied sharks moving ghost-like through the coral

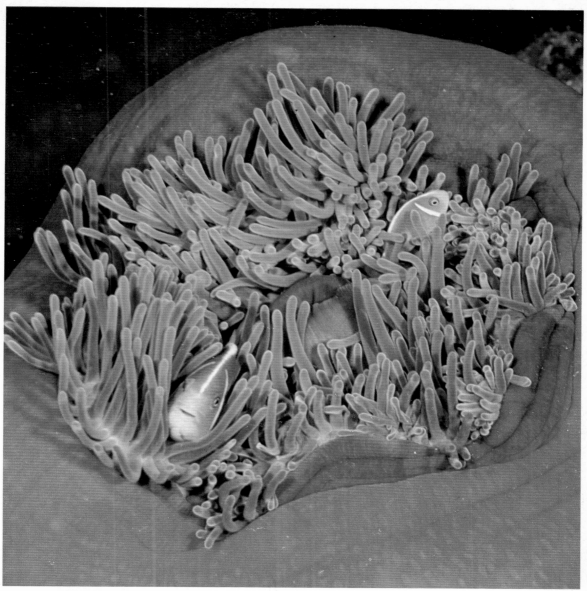

In the Indian and Pacific oceans, anemone clownfish (Amphiprion) are found exclusively in association with anemones, which provide them with a home and protection from enemies. For reasons still unknown, these fish are not stung by the deadly tentacles of their hosts. If taken away from its host anemone for several months and then returned to it, the clownfish is no longer safe from the tentacles. But gradually, by touching parts of its body to the anemone, the clownfish may regain its immunity.

Fishes of the coral-reef community, among the most brilliantly colored of living organisms, range in length from 15 to 20 centimeters. In a reef environment, more than 100 species may be represented within only a few cubic meters of water. Row 1: far left, *garibaldi* (Hypsypops rubicundus); left, *humuhumu* (Rhineacanthus aculeatus); right, *white-faced surgeonfish* (Acanthurus japonicus). Row 2: far left, *rock beauty* (Holacanthus tricolor); left, *fairy bass* (Microlabrichthys tuka); right, *blue-ringed angelfish* (Pomacanthus annularis). Row 3: far left, *emperor angelfish* (P. imperator); left, *harlequin wrasse* (Lienardella fasciatus); right, *blue surgeonfish* (Paracanthurus hepatus).

Barracudas (Sphyraena), above, *and lionfish* (Pterois), right, *are two of the reef's most-feared predators. Barracudas not only are dangerous to swimmers but may be poisonous even if cooked and eaten. Lionfishes possess many long fin-spines that can inject a painful, dangerous toxin.*

maze, seizing and tearing the hapless fish they encounter to bloody shreds. But on the reef, predators come in all sizes. The multitude of coral polyps are themselves tiny but very efficient predators.

The most interesting and active of the reef predators are the fishes. The many crevices and nooks among the corals furnish excellent hiding places from which they can dart out and seize their prey as it swims by. We may find jawfishes of the family Opisthognathidae, basslets of the family Serranidae, and the arrow blenny (*Lucayablennius zingaro*) skulking in reef caves, ever-watchful for victims.

The coral caves restrict the predators' movements and reduce their view of the area but, at the same time, offer protection from sharks and barracudas that are also on the hunt.

The great barracuda (*Sphyraena barracuda*) has been called the pike of the sea because of its appearance and its hunting behavior. It is long, lean, swift, and voracious like the freshwater pike, and its long snout is armed with cruel-looking teeth. Barracudas frequently hover around the edges of coral reefs waiting for prey fishes to appear. They remain nearly motionless, their caudal fin and ventral fins barely moving. Their coloration, light on the underside and dark on the upper side, also helps their concealment. When a school of anchovies swims nearby, the barracuda darts into the mass of small fish to attack a single individual and pays no attention to the others in the school. The anchovy eaten, the barracuda resumes its watchful wait, and the school swims off in panic, tightening its formation.

Some small predators shun the reef caves because they have other, more effective means of concealment and food-gathering. The spotted scorpion fish (*Scorpaena plumieri*) is festooned with fleshy appendages. With its natural camouflage, looking like an algae-covered rock, it waits until it can rush out to ambush its prey.

Anglerfishes and rays rest on the bottom among the coral heads to entrap small fishes and shrimps. The angler opens its mouth and dangles its fishing lure enticingly in front of its deadly jaws. The rays arch their broad "wings" to form a cave, and unwary prey swim into the opening, seeking refuge, only to be quickly gobbled up. The trumpetfish (*Aulostomus maculatus*), a long, slender fish, hangs vertically in the water among soft corals and subtly changes color to match its background. Virtually invisible, it snatches at tiny planktonic animals.

One predator that needs no camouflage or hiding place from which to launch an ambush is the lionfish (*Pterois*). Its striped decorations and long, feather-like fins make it one of the most striking and gaudy of the reef fishes. The most richly decorated of the lionfishes is *Pterois antennata*. Above its eyes, the lionfish displays plumed lances which are weapons with venom glands containing powerful toxins. The stiff fin rays that hold the elaborate dorsal fins erect are also part of the fish's venom-spined finery. Few predators dare attack the lionfish because of its poisonous defense. Bathers and skin divers who accidentally touch the lionfish have received painful, near-fatal wounds from the venom apparatus.

Other fishes with dangerous spines include: top, *crocodile fish* (Platycephalus); *and* bottom, *scorpion fish (Scorpaenidae). These bottom-dwelling fishes inflict painful, sometimes deadly wounds if trod upon.*

There are more than 500 known species of eels, including many morays (Muraenidae). They are active at night and retire to holes in the reef during the day. Most eels have a very sensitive sense of smell, especially such morays as: top, Gymnothorax; and bottom, Rhinomuraena. Even though some species grow to large size (2 meters long and as much as 30 kilograms), they almost never attack human beings without provocation.

The moray eel (*Gymnothorax*) is one of the most gruesome-looking and feared of the reef predators. By day the eels rest in coral crannies, ever alert for a small fish or other prey. At night they join the reef sharks in a grisly search for food. Sometimes a moray will drape itself over a convenient coral head like a wet rope. Its sharp teeth, revealed in what seems an evil grin, indicate its role as a predator. Although morays are not known to deliberately attack man, skin divers who foolishly tried to pull a moray from a coral crevice have found themselves trapped in its toothy grip. The lacerations on their hands and arms gave painful evidence of the power of the moray's bite.

Although the classic predator of the seas, the shark, has been discussed more fully in a previous chapter, the reef is the favorite hunting ground of certain sharks because of the abundance of fishes and other prey animals found there. White-tipped reef sharks (*Triaenodon obesus*), 1 to 2 meters long, patrol the calm waters of atoll lagoons, seeking food. Spear fishermen frequently attract sharks because of the blood and fluids oozing from fish they have speared. One diver, approached by a reef shark, abandoned his catch and beat a hasty retreat. From a distance he watched in amazement as a large moray eel snaked from a coral crevice and snatched the fish from the shark's jaws. Whirling round and round in the bloodstained water, the shark and moray struggled to possess the carcass. Finally the fish was torn apart, and shark and moray swam off, each with a large chunk of flesh in its mouth.

The productive reefs are regularly patrolled by other large sharks, including the deadly tiger shark (*Galeocerdo cuvieri*), which is a resident of both the Atlantic and Pacific oceans and may grow to 4 meters in length.

The Great Barrier Reef of Australia is well known for the large number of great white sharks (*Carcharodon carcharius*) cruising its sun-warmed waters. The other common names for this shark—white pointer, white death, and man-eater—clearly indicate its fearsome nature. Curiously, one of the most terrifying-looking members of the reef community is the least harmful. Skin divers probing among the coral blocks have sometimes looked up to see the sun blotted out by a huge shape "flying" through the water overhead. The diver's first reaction is fear, because the creature he sees is a giant manta ray, a relative of the shark and closely related to the skate, with a wingspread that may measure 5 to 8 meters from tip to tip. Some of these giants weigh nearly 1,500 kilograms. But the diver has little to fear from the manta—it is a plankton feeder.

Cleaning Stations
Although coral reefs may appear to be underwater jungles ruled by the law of "Kill or be killed," "Eat or be eaten," they also have "truce" areas. These are the cleaning stations, where predators and prey engage in ritualistic cleaning. The cleaners perform elaborate movements, including a sort of shivering, posturing, and displaying of fins or appendages to advertise to potential patrons

Above and left. *Sea snakes are
found in the Indian and Pacific
oceans as far east as Ecuador and
the Gulf of California. Typical of
several species, all venomous, is
the olive sea snake* (Aipysurus
laevis). *Sea snakes reach lengths
of 2 meters or more. Occasionally,
they must come to the surface to
breathe air, like marine turtles.
122–123. Mantas* (Manta
birostris) *are awesome creatures;
their wing tips span 7 to 8 meters,
and they weigh as much as 1,500
kilograms. They often bask at the
surface of the water, sometimes
turning somersaults and often
leaping out of the water and
falling back with a thunderous
splash. Mantas have tiny teeth
and feed on plankton.*

Some small fishes set up cleaning
stations and remove parasites
from larger species. Above. A
grasby (Petrometopon
cruentatus), in the Caribbean,
opens its mouth for cleaning by
the sharknose goby (Gobiosoma
evelynae). Right. A squirrelfish
(Holocentridae), in the Palau
Islands of the Pacific, positions
itself for cleaning by the wrasse
(Labroides bicolor). The large
fishes recognize the cleaners by
their color pattern and behavior.
The fishes being cleaned almost
never eat the cleaner, even though
cleaners often swim into the
mouths of their "clients."

that cleaning services are available. Tiny fishes and shrimps, ordinarily the prey of groupers, barracudas, and other reef fishes, pick over the bodies of the larger fishes to rid them of parasitic isopods and copepods. They flit around the fins, over the eyes, and even into the larger fishes' gaping mouths to get at parasites or bits of food caught between the teeth.

This cleaning serves a dual purpose. The large fishes are rid of debilitating parasites, and the cleaners feed on the organisms and food scraps they energetically remove from the hosts.

Death of the Reef

The solid and seemingly permanent appearance of the reef is an illusion. The reef is actually a battleground between long-term constructive forces and powerful destructive forces. Construction, of course, includes the natural growth of the reef as the polyps build their skeletons upward at the rate of about 2.5 centimeters each year. Calcareous algae add material and help bind the reef together. Violent ocean storms carry bottom materials, including sand and fragments of shells, onto the reef.

Crashing waves that tear away the reef during typhoons, hurricanes, and other mighty storms contribute to eventual destruction of the reef. Boring sponges, worms, and clams riddle the reef with holes that eventually weaken it. During extreme low tides, the polyps are baked by the sun or weakened and killed by the fresh water of rain.

Some devastation of coral reefs is part of the natural cycle of predation in the sea. For example, not too long ago, the world learned of the insidious destruction of the Great Barrier Reef of Australia. The culprit was a large, spiny sea star: the crown-of-thorns starfish (*Acanthaster planci*). Many thousands of these sea stars crawled over the reefs in a massive invasion—caused, it seems, by a population explosion. They devoured the delicate coral polyps and left many hectares of dead reef vulnerable to erosion, with no polyps to build and maintain the structure. In 1970, it was reported that the crown-of-thorns starfish had destroyed 8.1 percent of the Great Barrier Reef.

The enormous increase in sea stars was thought to be the result of a reduction in the animal's natural predator, the helmet shell (*Cassis*). This large conch with its massive shell would climb over the crown-of-thorns, apparently immune to the spines, smother it, and then in leisurely fashion, consume it. However, human shell collectors, eager to obtain specimens of the attractive helmet shell, had badly depleted its numbers. No longer threatened by the helmet shell, the sea star rapidly increased, to the detriment of the coral animals and the reef.

The most recent evidence suggests that the crown-of-thorns starfish has reached a population peak and has begun to decline in numbers as the result of disease, overcrowding, and environmental factors, such as minor changes in water temperature and fluctuation of currents. Research is continuing to evaluate the impact of the starfish. However, the scientists' task is not an easy one,

On a Tahitian reef, saber-toothed blennies (Aspidontus taeniatus) *induce larger fish to pose for cleaning but, instead of removing parasites, take bites out of the larger fishes. Some scientists believe the blennies mimic true cleaner fish—the wrasses—in color pattern and behavior, so they can approach and feed on other fishes.*

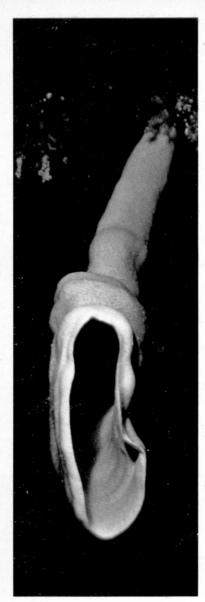

Sponges, such as Siphonochalina, *have generally porous bodies composed of fibrous material; they are usually found in colonies attached to underwater surfaces.*

because the reef is a dynamic entity that changes from day to day.

There is no doubt that the greatest damage to the coral reef community—and the actual death of the reefs—is caused by man's activities. Some of the damage is deliberate, while some results from indirect or remote activities.

Collecting of colorful reef fishes for the home aquarium trade reduces the numbers and varieties of reef organisms. Reef ecologists are greatly concerned that unrestrained netting of reef fishes by commercial aquarium collectors will result in all the evils of overfishing. Further, in their eagerness to take greatly desired specimens, many collectors break down living coral heads, thus contributing to the death of the reef. Thoughtlessly, large formations are loosened with crowbars, sledgehammers, and even dynamite.

Some tropical spear fishermen are using household bleach (sodium hypochlorite) to drive fishes out of holes in the reef. But the bleach also kills the coral polyps, kills or drives away fishes and invertebrates, and destroys sponges and tube-dwelling worms. Although such "bleach fishing" is illegal, enforcing the laws is very difficult. In most places the fisherman must be directly observed using the bleach before he can be arrested.

Reefs are "mined" for agricultural lime, cement, and road-building materials. Entire coral reefs and the associated communities have been smothered by silt. Forests that have been cut severely for timber, intensive agriculture for pineapple and sugarcane, road-building, and land clearing for housing tracts have all resulted in tons of soil being washed into the sea and over the reefs. Sewage from inhabited areas has poisoned the reefs. In many places, the dead and dying reefs have become overgrown with thick layers of green bubble alga (*Dictyosphaeria cavernosa*). Like some science fiction monster, this alga covers the once-vital reef in a hideous shroud as much as 35 centimeters thick. Nothing will grow underneath it.

Will such wasted reefs ever again become the vital, growing communities they once were, home to so many swirling, fanciful, and brightly colored fishes? Probably not. To their dismay, researchers have found that in some places in the Pacific Ocean, coral reefs killed 40 years ago still show no sign of recovery or regrowth.

Above. *Like* Callyspongia vaginalis, *most sponges are tubular. The walls of the tubes are porous, with tiny canals leading into a central cavity. By means of ciliary action, the cells that line the canals transport water into the central cavity and out of the hole at the top. As water passes through the sponge—sometimes called the "filter of the sea"—it removes food particles. Sponges are also homes for many marine organisms, such as shrimp and small fishes, which live in their central cavities.*

128–129. *Darwin considered the reef of Bora Bora in French Polynesia, which is separated from a volcanic island of the Leeward group by a wide lagoon, a classic example of the barrier reef. According to his theory, the reef marks the old coastline of the island, which the lagoon gradually encroached on as the island sank into the sea. Perhaps someday the island will completely disappear beneath the waves, leaving only a ring of living reef around a circular lagoon.*

The Open Sea

Long before the great exploratory voyages of seafarers like the Vikings, Columbus, Verrazano, and the ancients before them, the open sea remained a terrifying place, supposedly inhabited by monsters that could devour ships and men. And the vast uncharted expanses of sea were thought to lead to the edge of the world, where a vessel might plunge over the watery precipice into oblivion. But the early explorers found that the open sea was actually a vast emptiness, a monotonous watery realm of tossing waves and swells. Occasionally a great whale or school of porpoises might be sighted; or an island of flotsam, complete with trees, birds, insects, and small mammals might drift by the vessel like some ghostly mini-continent.

But other than these occasional sightings, no living things were to be seen in mid-ocean.

Indeed, learned men long considered the open ocean to be virtually a biological "desert." Then oceanographic expeditions began demonstrating that the open ocean was populated with a host of marine animals. Each ocean revealed unknown animals, different from those seen elsewhere. Some stretches of the open sea, of course, still seemed to support the old beliefs of a "sea of no living thing."

We know today that what the early explorers saw was a result of the influence of different ocean environments on the distribution and abundance of marine animals.

Each ocean and sea, and even parts of the oceans and seas, evolved plants and animals different from those in other parts of the world ocean. For example, tropical marine organisms cannot migrate between the Atlantic Ocean and the tropical Pacific Ocean. To do so, they would have to cross long stretches of cool water, which would be fatal. This was not always so, of course. In ages past, there was free passage between these waters, but as the Isthmus of Panama emerged as a land bridge between the continents of North and South America a relatively few millions of years ago, the seaway between the Caribbean and the tropical Pacific was cut off. The great mass of the African continent forms the same sort of barrier between the marine organisms of the tropical Atlantic, the Mediterranean, and the Indian Ocean.

Another factor affecting the distribution and abundance of marine animals is the availability of food, which is controlled, in part, by the amount of sunlight. In the open ocean, far from the coasts and at the uppermost zone, daytime illumination from the sun is too strong for the survival of most planktonic plants and animals. Only a few specialized organic forms are found in this zone. Since this zone may extend down nearly 20 meters, it is not surprising that the absence of plankton and the fishes that feed on them gave rise to tales of "dead seas."

The next vertical zone in the open ocean is that about 20 to 40 meters below the surface. In this, there is much photosynthetic activity, and the microscopic plants found here use the faint energy of the sun to produce food for marine grazers.

Below the photosynthetic zone, the water is populated by vast numbers of zooplankton that are transients, moving upward at night to graze on the meadows of phytoplankton and then downward again to escape the illumination of daylight. In the Sargasso Sea, hordes of tiny animals have been found at depths of about 100 to 150 meters during the day.

Zones of Productivity

Early oceanographers easily grasped the differences in vertical distribution and abundance of plankton, from the sunlit surface to the darkened depths. What they could not understand were the horizontal differences, that is, the scarcity of plankton and the absence of fishes across wide reaches of the sea. As research and exploratory ships traversed the open waters, nets of finest gauze that were towed through the upper meter or so of the ocean were often retrieved virtually without life forms. Such pockets of low productivity were found in the central portions of the northwest Atlantic Ocean. An especially "dead" pocket was found across the South Atlantic; others were found in the Indian Ocean between Australia and Madagascar and in the great central basin of the Pacific Ocean. Curiously, however, in the Pacific a "tongue" of productive waters was found along the Equator.

Despite large zones of low productivity, here and there the waters of the world ocean provide the conditions favorable to the growth of phytoplankton. Most of these areas of increased productivity are located along the edges of the great equatorial currents, where characteristic water turbulence stirs up vast amounts of nutrients from the bottom. These nutrients become available to the tiny plants, and the cycle of marine life is renewed. Fishermen on the high seas, knowing that large fishes can be found feeding on smaller fishes nurtured by the clouds of plankton in these enriched areas, have learned to take advantage of them.

Drifters of the Open Sea

The development of modern fishing gear and ingenious oceanographic research devices first revealed many of the organisms existing in the open sea; but others had been known for centuries. The leisurely progress of sailing vessels and the long days spent at the rail gazing out over the water gave earlier mariners many opportunities to observe organisms on the water's surface. One creature for which they might have felt a kinship is a relative of the jellyfish known as jack-sail-by-the-wind (*Velella velella*). This colonial animal, 6 centimeters long and 4 centimeters wide, resembles a miniature raft. On top of the raft is a triangular fin that acts as a sail to catch the wind. Below the raft hang the polyps and tentacles with which the organism captures and consumes its prey. Some years, the winds and currents carry armadas of the tiny blue rafts across the storm-swept Atlantic in miniature invasions that eventually wash up in windrows on the shores of northwest Europe.

These curious invaders often are accompanied by the planktonic snail (*Janthina janthina*), which uses a bubble raft to float its tiny shell (1 centimeter diameter) just

132–133. *Jack-sail-by-the-wind (Velella velella) has a gas-filled float with a triangular "sail" on top. Thus blown before the wind, it tows tentacles through the water and gathers whatever food it touches.*

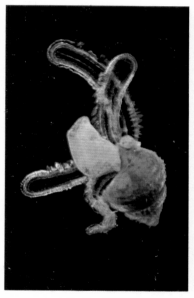

Plankton is a nursery for many organisms that are bottom-dwellers as adults, such as this veliger larvae, the characteristic larval form of marine snails. Adults shed their eggs and sperm into the water and give them no further attention. Drifting for weeks or months, the larvae disperse widely, reaching areas hundreds or thousands of kilometers distant from their parents. On encountering a suitable shoreline, or simply settling on the ocean floor, the larvae transform into bottom-dwelling miniatures of adults.

Janthina janthina *is a purple
floating sea snail that feeds on*
Velella velella. *Above.* Janthina
*floats upside-down from its bubble
raft, at Key Biscayne, Florida.
Right.* Janthina *feeds on a* Velella;
*it consumes the tentacles and body
but leaves the float untouched.*
Janthina *overcomes its prey by
secreting a concentrated purple
dye that renders the stinging cells
in the* Velella *tentacles harmless.*

below the surface. During an oceanographic expedition aboard the research vessel *Discovery II*, a scientist noticed large numbers of *Velella* and *Janthina* in the Atlantic about 800 kilometers west of Cape Finisterre, on the extreme northwest coast of Spain. Many *Velella* had small *Janthina* attached to their undersurface. The scientist put several of the animals in a shipboard aquarium and observed the snails browsing on the *Velella*, clearing the underside of its tentacles, and at the same time exuding a purple dye. The dye, he reasoned, was a substance to anesthetize the floater and render it helpless. Eventually, nothing was left of the *Velella* except the horny raft. Such rafts drifting on the open sea often become floats on which goose barnacles (*Lepas anatifera*) attach and grow.

Early sailors probably also saw fleets of the Portuguese man-of-war (*Physalia physalis*) drifting across the water surface like so many small balloons. This beautiful but deadly jellyfish has a large, gas-filled float about 35 centimeters long, with a crest that acts as a sail. Periodically, the colonial animal dips the crest into the sea to keep its membrane wet and flexible. Boatmen are often tempted to pick up the floating object, attracted by the colors of the float and crest. The delicate membrane is an iridescent blue that shades into mauve and pink at the top of the crest, and tinges of orange highlight the whole. But there is danger beneath the attractive float, for extending as much as 10 to 15 meters below and behind the float are deadly stinging tentacles capable of paralyzing large fish and drawing them up to the digestive apparatus for consumption. As the float drifts along, driven before the wind, its tentacles stream out behind, fishing for prey. Anyone coming in contact with the tentacles is stung by tiny darts, the nematocysts, which are like miniature hypodermic syringes filled with poison. The poison of the Portuguese man-of-war can produce sensations of stinging, burning, or numbness and result in extreme pain, nausea, and even, in certain individuals, shock.

Physalia is found throughout tropical waters but drifts toward the poles during the warm months. Although primarily creatures of far-offshore waters, many of these drifters, when sea and weather conditions are right, are blown up on the shores of northwestern Europe. The stranded specimens are still capable of stinging anyone who picks up these beautiful but lethal creatures.

"Sea Serpents"

Whales, porpoises, large sharks, and even the remains of giant squid probably inspired the widespread ancient tales of sea serpents. But no doubt some tales were also inspired by sighting, or finding the beached remains of, some of the other bizarre fishes of the open sea, such as the oarfish (*Regalecus glesne*). One of the rarest fishes in the sea, it is nevertheless worldwide in distribution, and specimens have been caught in the waters off both Europe and Japan. Most of what we know about it is based on skeletons or partially decomposed remains found on windswept strands.

A larger relative of Velella *is the Portuguese man-of-war* (Physalia), *whose tentacles may extend down into the water for many meters. When the tentacles contact a small fish or other food item, they contract and thereby draw the fish up to the digestive organs under the float.*

137. *At nesting time Green sea turtles* (Chelonia mydas) *migrate to certain stretches of sandy coast. There the males and females congregate offshore. As shown by this pair in Indonesia, mating occurs in the water and can be awkward, since both partners must keep their heads above water in order to breathe. The male grasps the female by means of a claw on each front flipper and also with his tail. The flipper claws sometimes cut deep notches in the bone of the front edge of the female's shell.*

About 50 species of flying fishes (Exocoetidae) *inhabit the warmer parts of the world ocean. "Gliding fishes" would be a better name, for they gain speed while still in the water and then leap out while rapidly vibrating their tails, which have elongated lower lobes. The added thrust of the tails aids the fishes in becoming airborne. Once aloft, they glide, with no apparent control, on their greatly enlarged pectoral and pelvic fins, which are merely extended and never flapped. Flying is these fishes' way of escaping predators, especially the fish-dolphin* (Coryphaena). *Their longer flights cover distances of hundreds of meters. Sometimes smaller individuals will emerge together from the water and, if caught by the right breeze, will seem to fly off in formation.*

The oarfish, which may grow to be 6 meters in length but only 30 centimeters thick, is flattened and ribbon-like and swims in undulating fashion like a snake. The colorful oarfish looks as though it were painted by a poster artist. Its glistening silvery body bears a continuous, brilliant scarlet dorsal fin. The fin atop the head is enlarged into an astonishing mane or crest; below, the scarlet pelvic fins are long and drawn out, with broadened tips like oar blades—hence the fish's name. In northern Europe the oarfish is called "king of the herrings" because of the erroneous belief that it swims with herring shoals.

Curiously, the Pacific coast Indians of North America have a similar belief about *Trachypterus altivelis*, a relative of the oarfish they call "king of the salmon." The alleged "king" is seen just before the salmon begin their spawning runs up the coastal rivers, and the Indians believe that if the king were killed, the salmon would abandon their run. (*Note:* This should not be confused with the well-known king salmon, an entirely different species, *Onchorhynchus tschawytscha.*)

Rivaling the oarfish in coloration, if not size and shape, is the opah, or moonfish (*Lampris regius*), another inhabitant of the mid-depths in the open sea, which grows to a more modest length of 1 to 2 meters. Its thin, compressed body is a half to 1 meter deep and tinted with brilliant, varied hues. The upper parts are a dark steel-blue, shading downward into a green with silvery, purple, gold, and lilac luster. Its belly is rosy, fins are vermilion, and the whole body is speckled with silvery and milk-white spots. Known throughout both the Atlantic and the Pacific oceans, the opah actively feeds on squids and small fishes, which it generally captures at depths of 90 to 180 meters. Specimens of this radiant species have been collected in waters off Madeira, Scandinavia, the British Isles, Iceland, Newfoundland, Nova Scotia, and Cuba and in the Gulf of Mexico.

Oceangoing Reptiles

Legends and myths of ferocious sea serpents abound in all cultures bordering the world ocean. But the only genuine serpents or reptiles found in marine waters are sea turtles and sea snakes. With the exception of the sea snakes, which rival cobras in venom, marine reptiles are relatively harmless. They are now few in number, but this was not always so.

In the dim reaches of the planet's early history, during the Mesozoic—about 130 million years ago—the seas abounded in large and fearsome reptiles. Basically there were three groups: the whale-like ichthyosaurs, the long-necked plesiosaurs, and the large sea lizards, the mosasaurs. This was the Age of Reptiles, when "thunder lizards," the dinosaurs, ruled supreme on the land. But, like the dinosaurs, the great marine reptiles became extinct at the end of the Cretaceous, about 50 million years ago. The reptiles of our present-day seas are little more than remnants from their glorious past. The sea turtles and sea snakes are tropical or subtropical in distribution, but many are carried far into the open sea by such "rivers of the sea" as the Gulf Stream.

Green sea turtles must lay their eggs out of water. Slow and awkward on land, they are helpless against human hunters, and their mortality rate is high. Only the females come ashore at laying time—every 12 days or so, until three to five clutches of eggs are laid. Above. With her hind legs, the female digs a flask--shaped hole well above the high-tide zone and deposits her clutch. She covers the hole with sand and conceals its location by kicking loose sand around the nest. She then moves slowly back to the sea. The nest contains about 100 eggs, which hatch in approximately 60 days if not dug up by predators. Right. The hatchlings cooperate in digging themselves out of their chamber, then move in a loose group toward the sea.

Four species of sea turtles from the Caribbean and nearby tropical waters have been recorded in the waters off the northwest coast of Europe. Frequently they are seen far out in the ocean, hundreds of kilometers from land. These large, clumsy-looking turtles, some with shells a meter or more across, paddle steadily along, their great bodies encrusted with barnacles, hydroids, algae, and other marine hitchhikers. Occasionally the underside of a turtle may carry one or two *Remora*, going along for the ride. These strange fishes, 13 to 60 centimeters long, have on top of their head a curious flattened sucking apparatus with which they attach themselves to turtles, whales, porpoises, or even ships, but especially to sharks. Hence they are often called "shark suckers."

A rare and valuable species among the turtle emigrants is the hawksbill turtle (*Eretmochelys imbricata*), whose shell—the "tortoiseshell" of commerce—is made into expensive combs, mirror backs, and jewelry. The horny plates on the top shell of this species overlap like slates on the roof of a house. This is a large species, often 1 meter in diameter, with a head and beak very much like those of a bird of prey. The tropical American variety of hawksbill turtle is but one of several around the world. These lumbering giants occur in the Mediterranean and in the warm waters of the Indo-Pacific realm. In the eastern Pacific they are found along the coast from Baja California to Peru, while in the western Pacific they occur from southern Japan to Australia.

In certain years, the vagaries of the Gulf Stream and North Atlantic Current, combined with forceful southwest winds, bring two similar-looking turtles into the waters off the northwest coast of Europe, far from their customary tropical and subtropical nesting areas. One is the common loggerhead (*Caretta caretta*), and the other is the ridley (*Lepidochelys kempii*). Both may grow to be more than a meter in diameter. Occasionally, loggerheads get caught up in other ocean currents, and surprised fishermen and beachcombers find them from Nova Scotia to the Shetland Islands and southward to Uruguay and the Cape of Good Hope. Though widely distributed in the Indian Ocean, the species are not common there. In the Pacific, loggerheads beach themselves to lay their eggs in the warm sands of myriad islands from the northern part of the Sea of Japan to southern Australia.

By far the strangest-looking and largest of these sea turtles is the leatherback, or luth (*Dermochelys coriacea*). This behemoth lacks the top shell of horny plates but is covered with a thickened, ridged leathery skin in which is embedded a mosaic of many small bones. Because of its immense size, no predators would dare attack this animal, except possibly for a very large, carnivorous shark or the killer whale. It frequently grows to nearly 2 meters in diameter and weighs up to 500 kilograms. In a large adult 1.7 meters in diameter, the enormous, wing-like front flippers have a spread of nearly 3 meters!

The leatherback is a solitary wanderer, at home in all temperate and tropical ocean waters. Its worldwide travels suggest the likelihood that, having spent the summer

140–141. *Hawksbill turtles* (Eretmochelys imbricata) *are the smallest of the sea turtles. The carapace, or top shell, of a very large adult may reach 75 centimeters in length. The narrow, strongly hooked beak is used to catch and kill fishes, crustaceans, and mollusks. Hawksbills are avidly hunted for the beautiful brown-and-yellow marbled plates of their carapaces—the genuine "tortoise-shell" used in combs and jewelry. One species inhabits the Indian Ocean; another, the warm waters of the western Atlantic. Hawksbills are thought to be closely related to the herbivorous green sea turtles.*

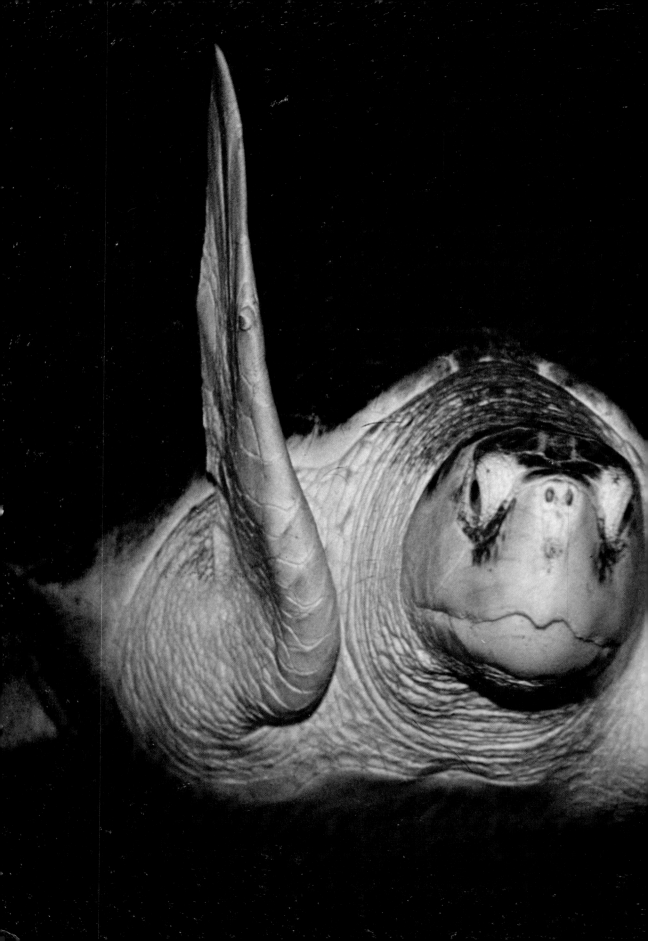

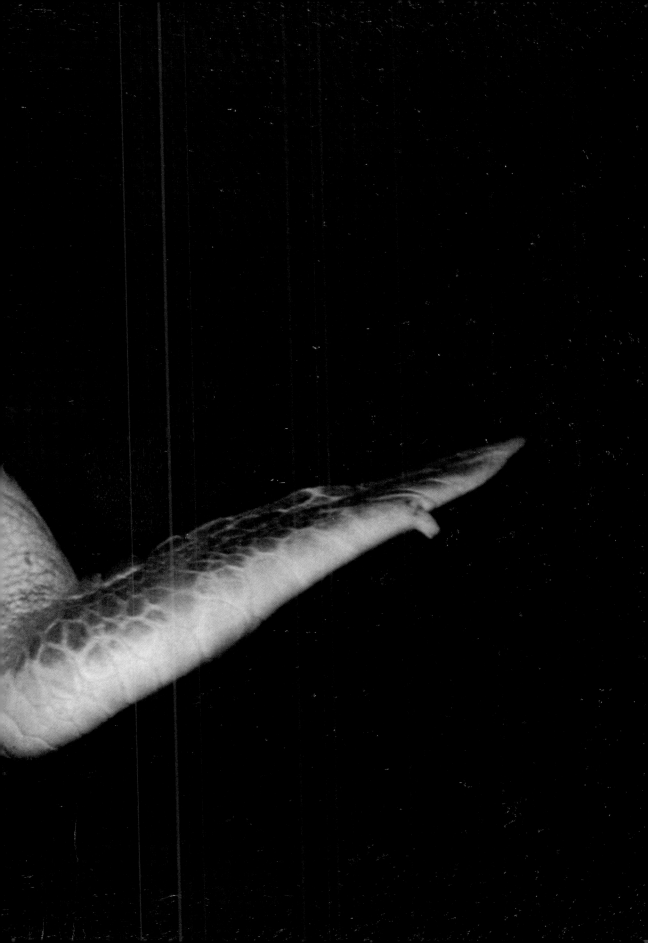

cruising off the rugged coast of Norway, a male leatherback may wend his way to Venezuela to find a mate that could have arrived from a river mouth in Uruguay.

Although adult sea turtles have little to fear from predators, some are themselves efficient predators, attacking and consuming fishes and a variety of invertebrates. Curiously, several turtles feed on the drifting, venomous jellyfish, including the dangerous Portuguese man-of-war. Sea turtles have been seen drifting along in the currents nibbling at the jellyfish's trailing tentacles and gas-filled floats; they are amazingly able to consume the jellyfish with no ill effects from the stinging nematocysts of its tentacles. Other turtles, such as the green turtle (*Chelonia mydas*), are vegetarians as adults and graze on turtle grass.

Like so many other dwellers of the world ocean, the sea turtles are in trouble everywhere, hunted ruthlessly for their eggs, meat, and shells. What human hunters don't kill, dogs, rats, snakes, and birds of prey may, so that the populations of these turtles are in serious decline. Indeed, except for the lands around the northern part of the Arabian Sea, whose sparse inhabitants are forbidden by custom to eat turtles or their eggs, the sea turtles are being slaughtered to extinction.

Sea snakes are true reptiles that have become specially adapted to a life in the sea. Their nostrils have valves that close when the animals submerge. The scales which are usually so prominent on land snakes and which help them move across the ground are very small or entirely absent on sea snakes. The tail, instead of tapering to a thin point, is flattened, making the sea snake a more efficient swimmer.

Most sea snakes are less than a meter long, but a few species get to be nearly 3 meters long. They are handsome-looking creatures with banded patterns on the skin; probably the most colorful is the yellow-bellied sea snake (*Pelamis platurus*), showing a rich black on top and bright yellow beneath.

Sea snakes are mostly coastal animals and are found throughout the tropical Pacific and Indian oceans, from the Persian Gulf to the Yellow Sea and Japan and to northern New Zealand.

The yellow-bellied sea snake, however, is a creature of the open sea, and individuals have been sighted actively swimming, with their characteristic sinuous motion, hundreds of kilometers from land. This species is widely distributed from the east coast of Africa throughout the Indo-Pacific to the west coast of the Americas (in the Gulf of California and south to Ecuador).

Anchovies, small herrings, eels, and mullet are the principal prey of these fish-eating reptiles. Although they try to avoid man, they become aggressive when annoyed and will actively pursue and attack whatever annoys them. Divers irritating sea snakes have found, to their horror, that an apparently fleeing sea snake can suddenly turn and sink its fangs into an arm or leg. Sea snakes are highly poisonous, with venom even more lethal than the cobra's. What may have started as an underwater frolic can have a fatal outcome.

Mystery of the Long-distance Swimmers

The waters of the open sea are crisscrossed by the invisible tracks of organisms that are long-distance swimmers. Much of what we know about such swimmers is gained from the work of fishery scientists, who place marks or tags on hundreds of individual animals of a particular species. Eventually a few of these marked individuals are again located. Thus, we know when and where an animal was tagged and when and where it was caught again, but we know nothing of where it was in between. It is rather like having the birth and death certificates of a man and trying to reconstruct his life history; nevertheless, such scientific researchers have collected some valuable information.

The technical reports issued by fishery scientists have data, for example, about the bluefin tuna (*Thunnus thynnus*) of the Pacific that was caught in the Sea of Japan two years after it had been tagged off California. Biologists calculated it had traveled a minimum of 6,980 kilometers, at an average speed of 11 kilometers per day. And there are data about bluefin tuna tagged off the southwest coast of the United States that were recaptured by commercial fishermen off Norway and Spain. But data about two astonishing long-distance swimmers—one a bird and one a fish—came not from tagging studies but from very painstaking research that rivals the intricate deductions of Sherlock Holmes.

The story of this formidable bird, the now-extinct great auk (*Pinguinus impennis*), has been pieced together from the observations of cod fishermen on the Grand Banks of Newfoundland, from museum specimens, and from writings of nineteenth-century naturalists. The great auk was very much an underwater bird. The story of its mindless slaughter to extinction has been told and retold as an example of human greed and total disregard for a valuable, interesting marine resource.

The great auk, a flightless bird that stood about 75 centimeters tall and had a robust body, resembled present-day penguins. Like them it was awkward on land but graceful and marvelously efficient in the water. Each year, great flocks of the birds swam nearly 5,000 kilometers from their wintering grounds on the Outer Banks of North Carolina to nesting sites on the rocky islands around Iceland, Greenland, and Newfoundland. During its feeding forays it would descend 60 meters underwater and swim for nearly a kilometer in search of herring, crabs, and other prey. Because the auk was so abundant, cod fishermen from Spain, Portugal, France, Germany, and Great Britain, on their way to the Grand Bank, stopped at the birds' barren nesting grounds to kill them and feast on fresh auk meat and to salt down their flesh for bait. Later, trade in auk flesh and feathers flourished in Europe.

The slaughter of the hapless birds was carried out efficiently and ruthlessly. The fishermen herded them together on the islands, clubbed them to death, and loaded the carcasses into boats. They also shot them with scatter guns loaded with bits of scrap iron, old nails, segments of chain, and lead balls. And, in an ugly

The razor-billed auk (Alca torda) *is a living relative of the extinct and flightless great auk. The razor-bill can fly both in the air and underwater. It feeds primarily on fishes, which it catches by swimming underwater with flying movements of its wings. In winter razor-bills inhabit the cold, offshore waters of the North Atlantic; in the spring they migrate even farther north to rocky coastal islands, where they breed in colonies on cliffs and crags. Each female lays a single egg on a bare rock and then hides it in a hole or crevice.*

inspiration of destruction, the birds were herded toward the fishing boats floating in the surf so that their crews would not have far to carry the carcasses. Then the auks were forced to walk a plank from shore set across the gunwales, where sailors waiting with clubs crushed their skulls and tumbled the carcasses into the boat.

No population of animals can endure such intense exploitation for very long. Since the auk mated for life and the females laid only one egg each year, reproduction could not keep up with the numbers killed. Soon the flocks of millions became flocks of thousands, and finally it became evident the auk was very scarce. Museums and private collectors began a desperate race to get skins of the great auk before the birds disappeared from the seas. On June 3, 1844, the last-known great auk was killed by collectors on Eldey Island, 16 kilometers west of Iceland.

The story of the long-distance fish represents a no less painstaking piecing together of information, but has a happier ending. It concerns the common eel (*Anguilla anguilla* of Europe; *A. rostrata* of North America).

For untold generations, the eel has been eagerly sought as a food fish and has been the subject of myths, legends, and old wives' tales. The eel was thought to be a relative of land snakes, or at least able to breathe air and travel great distances over the land. The appearance of eels in a stream or pond with no evidence of eggs or fry led to a belief that it somehow sprang full-blown from the mud through spontaneous generation. Simply stated, the life cycle of the eel, particularly the early life history, was largely a blank. Then, one breakthrough shed some light on the creature but, at the same time, deepened the mystery.

Biologists had routinely collected a small, transparent, leaf-like marine organism in the northeast Atlantic and the Mediterranean Sea and had named it *Leptocephalus brevirostris*—"thin-headed animal with a short nose." It was considered simply another of the sea's myriad dwellers until some inquiring Italian biologists collected a few and kept them in an aquarium tank in Messina, Sicily. In astonishment they watched the strange creatures miraculously transformed into common eels. The experiment, repeated time and again, was hailed as the solution to part of the mystery of the fish. But the question remained: Where did the eels spawn?

The work of the Danish marine scientist Johannes Schmidt provided the answer to that question. Beginning in 1904, he collected many leptocephali in his plankton nets while on an expedition west of the Faeroes. Determined to find the source of the tiny organisms, he took his research vessels farther and farther west, systematically searching the water with his fine-mesh nets. Finally, he found the smallest of what were now known to be eel larvae over the deep water of the Sargasso Sea, about 950 kilometers southeast of Bermuda. It seemed incredible, but the adult eels, after spending 5 to 7 years in the rivers, canals, and ponds of Europe, made a spawning pilgrimage of more than 3,000 kilometers through the open sea to lay their eggs in

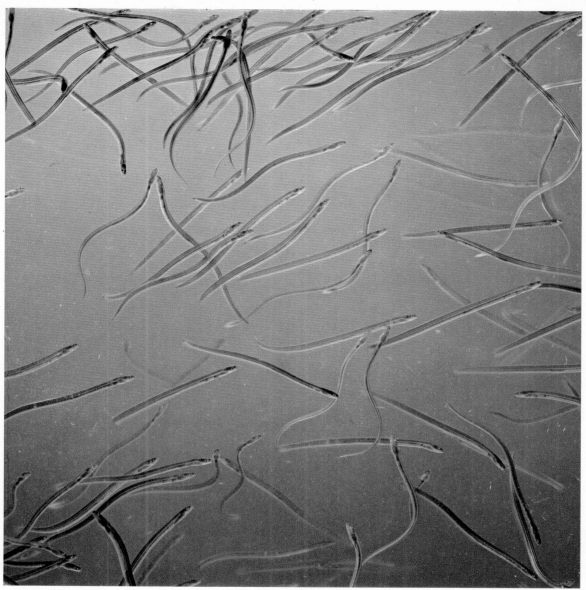

Above. *Elvers are young freshwater eels* (Anguilla) *that migrate upstream from the ocean. They remain in fresh water for several years before returning to the ocean as adults, ready for spawning. No eel eggs have ever been found—only small eels.*
Left. *The smallest of these are larvae, found drifting in the surface waters of the Sargasso Sea. The eel larva has a special name, Leptocephalus, because when first discovered it was thought to be an adult fish different from an eel. Leptocephali of freshwater eels spend two to three years drifting as plankton before they near shore and are transformed into elvers.*

Above, *Sailfish* (Istiophorus) *and, right,* marlin (Makaira) *are resplendent animals seldom seen except at the end of a hooked and baited line. Reaching a length of 4½ meters and a weight of 550 kilograms, marlins attain a maximum size only 50 percent longer than sailfish but may weigh five times as much.*

the abyssal blackness, some 1,800 meters below the water surface—and there to die. Curiously, the eels of North America make a similar journey to the Sargasso Sea to spawn. Eventually, by some strange genetic programming, the resulting larvae (still called *Leptocephalus*, although no longer a separate species) separate themselves. One species drifts off to European waters to metamorphose and mature, and the other migrates to east-coast North American waters.

Swordsmen of the Sea

Of all the fishes of the open sea, probably none are more impressive and awe-inspiring than the billfishes. Most are long and slender and bear a sword—simply an elongation of the upper jaw—that is slim and round or slightly oval in cross section. This sword is generally used to stun and kill prey, though occasionally it is used to impale other large fishes, including sharks and whales, and sometimes even pierces ships.

A large billfish found in the Atlantic Ocean is the Atlantic sailfish (*Istiophorus platypterus*), which has an enormous cobalt-blue dorsal fin and a long, slender spear or sword. Its average length is about 2 meters, but individuals as much as 3.5 meters long have been caught. The heaviest Atlantic sailfish weighed 60 kilograms, but the average weight is 18 kilograms. An active predator, the fish consumes great quantities of squid, octopus, paper nautilus, mackerel, and even small tuna.

In the Pacific Ocean, by far the largest of the several species of billfishes is the blue marlin (*Makaira nigricans*). The species was long thought to belong exclusively to the Pacific, but painstaking research and analysis of catches by long-line fishermen and anglers have revealed that it also occurs in the Atlantic and probably is worldwide in distribution. Some long-line catches have revealed individuals weighing more than 900 kilograms. Curiously, these very large specimens are females; males usually weigh only 90 to 140 kilograms. Tuna, swordfish, and squid are actively preyed on by the blue marlin.

The warm waters of the Indian and Pacific oceans are home to a giant among billfishes, the black marlin (*Makaira indica*). This great fish may measure 4.5 meters long, and specimens weighing 900 kilograms are common. Occasionally black marlin swim around the Cape of Good Hope and enter the South Atlantic. Thus it was that one weighing 500 kilograms was caught off Capetown, Republic of South Africa. Oddly, this giant fish feeds on the Portuguese man-of-war and apparently swallows its stinging tentacles without harm. It also feeds on squid and tuna; and one specimen was found to have a tuna weighing 73 kilograms in its stomach.

Sharing the Atlantic Ocean with the billfishes are several species of tunas. Some 1,200 kilometers off the west coast of Africa, in the Gulf of Guinea, the broad ocean area formed in the great bend of the continent, schools of bigeye tuna (*Thunnus obesus*), skipjack (*Euthynnus pelamis*), and yellowfin tuna (*Thunnus albacares*) feed and grow fat. Swiftly pursuing anchovies, flying fishes, and a host of small prey, the tunas represent the top of the

complex food pyramid nurtured by the elements pouring
into the West African gulf from great rivers like the Niger
and the Congo.

Several Atlantic species of tunas also occur in the
Pacific Ocean, including the bigeye tuna, yellowfin tuna,
skipjack, and albacore. All are actively sought with long
lines and purse seines, and a few are thought to be
overfished, although international management
schemes are in force in an attempt to conserve their
stocks.

The billfishes and the tunas have a type of protective
coloration, a sort of camouflage, that is common to many
fishes living between the surface and the depths. The upper
parts of their bodies generally are very dark, usually dark
blue, while their undersides are very light, silvery or even
white. Thus, swimming in the depths, such a fish will be
nearly invisible to a predator swimming above it. The dark
hues tend to blend into the dark blue-green of the great
depths. If the predator is swimming below, the silver or
white tones of the fish above would tend to blend into the
light colors of the surface waters and the sky overhead.

Homing of the Salmon
The broad stretches of the Pacific Ocean are well populated
with billfishes and tunas, but the northern portion of
this vast body of water is best known for its schools
of salmon. Six species occur in the North Pacific, five that
spawn in streams and rivers of the west coast of
North America and one, the cherry salmon (*Oncorhynchus
masou*), that spawns in Asian waters. All the Pacific
salmon belong to the genus *Oncorhynchus*, and all vary
greatly in size and appearance, though sharing many
similarities in life history.

Of the five North American species, the smallest is the
pink, or humpback, salmon (*O. gorbuscha*), whose
rich silvery scales are marked with small dark spots on
top. A mature pink salmon may weigh between 1 and 2
kilograms, and an exceptionally large one up to 4.5 kilograms. The largest is the chinook, or king, salmon
(*O. tschawytscha*), whose light-green upper scales are
marked with irregular black spots, contrasting with
its white underside. A very large specimen may weigh
56 kilograms.

There is considerable intermingling among the varied
populations of salmon in their feeding grounds in the Bering
Sea; some species have made migrations of more than
4,000 kilometers from their spawning ground. In the chill
subarctic waters, the salmon feed voraciously on the
abundant crustaceans, especially euphausiids, and on small
fishes, to become sleek and fat.

Depending on the species, salmon spend from 2 to 7 or 8
years at sea, growing in size and gaining sexual maturity.
When maturity is reached, the populations begin the long
journey to the stream or river where each was hatched.
Science does not yet know how salmon guide themselves
over the trackless wastes of the sea back to their home
stream. They may orient themselves to temperature
gradients; respond to the angle of the sun; or, according to
the most popular theory, detect minute "odors" from the

*Sockeye salmon
(Oncorhynchus nerka) are great
travelers, migrating far upstream
to spawn and then die. Center.
Their eggs are laid in the fall.
Bottom. Within eggs that are
several weeks old, the young fish
begin to take shape. After
hatching in the spring, the young
migrate downstream and
eventually enter the ocean.
Apparently retaining some
memory of the small stream where
they hatched, they return as
adults, after having lived two
years or so in the ocean, to the
same stream to spawn.*

Many dolphins are strikingly patterned: above, the spinner dolphin (Stenella longirostrius); right, Commerson's dolphin (Cephalorhynchus commersonii), from the Straits of Magellan. Commerson's dolphin, also called the piebald porpoise, is one of the smallest cetaceans—the group that includes dolphins and whales. The adult, less than 2 meters long, is found only in the cold oceans of the Southern Hemisphere.

stream carried in the fresh water as it mingles with the salt water of the Pacific.

As the salmon approach the home stream to make their final journey upstream to spawn, their outward appearance undergoes a mysterious change. The females become swollen with the eggs ripening in their bodies; the males develop a tremendous hooked jaw, and many do not feed and become emaciated, further increasing their grotesqueness. A few species take on bizarre coloration, becoming red with dark streaks. Leaping over waterfalls and swimming through swiftly running rapids, the salmon finally reach the site where they will deposit and fertilize their eggs. In a final flurry of excitement, the fertilized eggs are covered with gravel by the parents, now exhausted and scarred by the incredible rigors of the spawning journey. Feebly fanning their tail fins, the spent adults now drift in the stream they so recently fought against so strenuously and die, their life's work concluded.

Mammals of the Open Sea

Some marine animals look very little different from their land counterparts. Certainly this is true of the sea otter and the polar bear. Other marine mammals, however, have taken the great move forward and adapted perfectly to their watery existence. The seals and walrus have adapted well, but not nearly so well as the porpoises and whales. Seals and walrus must return to land often, to bask in the tepid warmth of the arctic sun or to give birth to their young. But porpoises and whales never leave the cold wet embrace of the sea except by accident, and then they die.

Porpoises and Dolphins

The ancients knew the porpoises and their kin, the dolphins, very well. Aristotle studied the playful beasts as they frolicked through the warm Mediterranean and Aegean waves. He concluded rightly that they were warm-blooded mammals, not fish. Some ancient peoples believed these animals held the souls of seafarers who had drowned at sea.

The ancient Greeks told tales of dolphins that approached bathers and allowed them to stroke their hides and play with them. Some dolphins were said to allow young boys to ride on their backs, and representations of a boy riding on a dolphin are common in early Greek culture. Such marvelous, seemingly implausible stories have been confirmed in recent years. One report comes from the New Zealand seaside town of Opononi, where a dolphin was said to swim among bathers, allow them to stroke it, and even permit a 13-year-old girl to ride on its back.

The dolphin was a frequent visitor to the harbor for several years but then disappeared. Nobody knows what happened to it, but there was talk along the waterfront that a drunken sailor had senselessly shot the gentle beast.

The two species familiar to the ancient Greeks and still well known in the same waters are the common porpoise (*Phocaena phocaena*) and the bottle-nosed dolphin (*Tursiops truncatus*). We now know, of course, that both

152–153. *The common dolphin* (Delphinus delphis) *is cosmopolitan; it is found in temperate and warm seas and occasionally even in fresh water. One of the swiftest cetaceans, it may reach speeds of 25 knots. Common dolphins usually travel in groups of 20 or more individuals—and on occasion even several hundred.*

Above. *The only truly white whale, the beluga* (Delphinapterus leucas) *sometimes travels in populations of more than 100, but these are really aggregations of smaller groups of no more than 10 individuals. The beluga is unusually supple for a whale and can swim backward by sculling with its tail. It lives in Arctic seas year-round. If trapped by the ice, it can sometimes break through by ramming with its head. Belugas have also been called "sea canaries" because of their trilling song, which can be heard above water.*

Right. *The blue whale* (Balaenoptera musculus) *is the largest mammal ever known. Formerly widespread in temperate and cold waters of both hemispheres, this giant is now a vanishing species.*

species are found in other parts of the world ocean, where they feed on fish and squid and seek out ships underway to ride the bow waves. In the Pacific Ocean the spotted porpoise and spinner dolphin (*Stenella*) actively swim with schools of yellowfin tuna (*Thunnus albacares*). Perhaps the porpoises seek the same prey species as the tunas, or perhaps they associate with the fishes simply because they like to. Whatever the reason, this proximity often results in the destruction of the mammals. Fishermen, knowing of this association and the likelihood of schools of tuna in the blue waters below, have learned to look for schools of porpoises. When nets are set out around the tuna schools, some porpoises are killed.

It is an exciting experience to see dolphins riding the bow waves of a vessel at sea—an experience already described by Greek sailors nearly 3,000 years ago. On one occasion I saw a trio of these animals riding the bow waves of our research vessel, easily keeping pace with our speed of 12 knots. They move effortlessly, and it is said they are able to swim in bursts of up to 25 knots, obviously having learned to take advantage of the hydrodynamics that operate as the bow of a ship cleaves the sea. Soon the sea virtually erupted with the graceful animals from horizon to horizon. Some took turns riding the bow waves; others merely swam along in the arcing movement called "porpoising." Still others engaged in a marvelous ballet, leaping 3 meters or more into the air and sometimes twisting as they rose. Some returned smoothly to the water, while others fell back on their sides, causing a great splash. It looked as if the dolphins were just having great fun, leaping and splashing like children playing in a pond; but as marine scientists, our judgment told us they were also probably trying to dislodge parasites from their skin.

Mighty Leviathan

While the porpoises are killed indirectly as the result of man's activities, their larger relatives, the great whales, have been deliberately annihilated. The worst example of this destruction can be seen in the depleted numbers of the blue whale (*Balaenoptera musculus*), the largest mammal the world has ever known. The massive size of the blue whale, which may reach a length of 30 meters and weigh 136 tons, surpasses that of the dinosaurs that once thundered across the land. Ruthlessly hunted by man as a source of oil, the blue whale, like most of the great whales, appears well along the irreversible path to extinction. Curiously, this leviathan—in another paradox of form and temperament—feeds on krill (*Euphausia superba*), a planktonic shrimp-like crustacean only 5 centimeters long; a whale consumes 2 to 3 tons at one feeding. During the period of their former abundance, blue whales swam freely in the coastal seas off Iceland, Alaska, Japan, Kamchatka, Mexico, Chile, California, and South Africa. Today, the remnants of the once-vast pods of blue whales are found mostly in the ice-rimmed waters around Antarctica.

The blue whale and its kind are called baleen whales—from the baleen, or whalebone, in their mouths, which strains

156–157. *The killer whale (Orcinus orca) is the largest member of the dolphin family (Delphinidae). Although most common in the Arctic and Antarctic regions, killer whales are found in all oceans. Hunting and attacking in packs, they easily overcome the larger but relatively defenseless baleen whales. They more commonly prey on seals, dolphins, fishes, and cephalopods. Sometimes they dislodge seals and penguins from ice floes by ramming the ice from below. The approach of a pack of killer whales usually panics other marine mammals. When menaced, gray whales are said to float motionless, belly up, if there is no nearby shallow water into which they can escape.*

Some whales undertake seasonal migrations, and one of the most interesting is that of the California gray whale (Eschrichtius robustus). Unlike migratory birds that breed at their summer grounds, gray whales breed at their winter home, in the warm and shallow bays along the Pacific coast of Baja California (Mexico). While migrating between Mexico and their summer feeding range in the Bering Sea, these baleen whales follow the coast. Whales appear to use visual orientation during their travels, for they frequently thrust their heads out of water to look around—a behavior pattern called "spy-hopping."

out the tiny food organisms in the thousands of liters of water it takes in during feeding. Another great whale, the sperm whale (*Physeter catodon*), is called a toothed whale because its powerful lower jaw is fitted with strong, conical teeth. Not as large as the blue whale, the sperm whale is nevertheless an awesome creature, reaching a length of 21 meters and weighing as much as 60 tons.

Although the blue whale is the largest of its kind, it is the sperm whale that is the most typical. This is the whale seen on desk ornaments, carvings, and prints at seaside resorts. It is a large animal—some measure nearly 20 meters long—with a huge, nearly square head that makes up about one-third of its mass. Like blubber from other members of the cetacean clan, sperm whale blubber can be rendered into oil of good quality. But the true commercial treasure lies deep in its massive head in the form of a special oil—spermaceti—used to lubricate watches and delicate machinery and also yielding a wax of highest quality for fine candles.

The whale actively seeks out the giant squids as prey. Men butchering sperm whales aboard factory ships marvel at the number of circular scars marking the animal's dark skin. Some scars are only 1 or 2 centimeters across, minor souvenirs of attacks by bloodsucking lampreys. Other scars, however, are nearly the size of dinner plates; these wounds are made by the sucking disks and hawk-like beaks of great squids as they ferociously struggle to free themselves from the whale's crushing jaws in the black abyss. Usually their struggles are in vain, and the whale takes its catch to the surface, where it bites off and swallows chunks of flesh the size of a football.

Although sperm whales are solitary hunters, the members of a "family" group have a social life. The head of the clan is a patriarch bull; he is mate to the older females and father of their young. A family may number 25 or so individuals, and it is the bull's duty to protect them from predatory sharks and killer whales. But the bull is virtually powerless against one predator: man and his harpoon guns and killer ships. As American marine biologist Victor B. Scheffer has pointed out, at the present rate of slaughter it will not be long ". . . before whales are remembered only from fading photographs and flickering videotapes."

Birth Underwater

Bearing their young alive is one of the characteristics that set mammals apart from other animals, and this is no different for marine mammals. Scientists have witnessed the actual birth of dolphins and whales, viewing them through glass observation ports in the tanks of "oceanariums." But except for the accidental emergence of a dead calf from its dead mother aboard a whale processing ship, no one has witnessed the live birth of a whale in its natural environment. What we can surmise about the process is based on what is known about dolphins, what is observed during commercial whaling operations, and what is known about the birth process in other mammals.

Birth is a major event in the sexual cycle of the great

Feeding behavior of baleen whales is related to the structure of the head, mouth, and tongue as well as the baleen. Top. Baleen plates, hanging from the sides of the palate in the southern right whale (Eubalaena australis), *resemble the teeth of a comb. The plates are about 2 meters long and fold away when the mouth is closed. Two rows of plates, one on either side, do not meet at the front. With its mouth open, the whale swims forward through swarms of krill. The food-laden water enters the mouth, the huge tongue blocks the passage to the throat, and the water exits through the baleen plates, which filter out the krill. Center. When enough krill has accumulated, the whale closes its mouth, dives, and swallows underwater, as this northern right whale* (Eubalaena glacialis) *prepares to do. Bottom. The diving humpback whale* (Megaptera novaeangliae) *takes big gulps instead of skimming, then swallows its filtered food by turning on its back, often with part of its head still above water.*

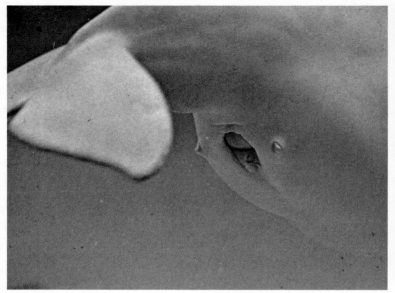

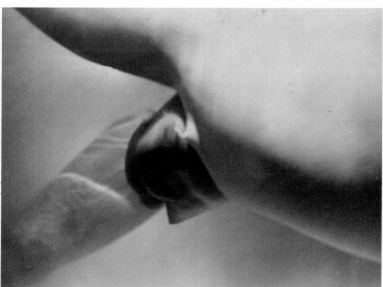

The birth of air-breathing creatures into a watery world is truly miraculous. Unlike land mammals, which are usually born headfirst, cetaceans often come out flukes foremost. Perhaps this position prevents the young from drowning during the approximately 20 minutes of the birth process.

The birth of a beluga whale begins with the opening of the birth canal. On either side of the canal are the teats, usually concealed in pockets in the skin. As the head is expelled, the mother twists sharply to the side, thus breaking the short umbilicus. The mother then nudges the infant to the water surface for its first breath. In some species, another adult female stays with the mother and assists in getting the young one to the surface. The baby's flukes and dorsal fin are folded and limp at birth; it swims weakly, but nevertheless must swim if it is to survive.

Right. *An adult male observes the activities of infant and female whales.*

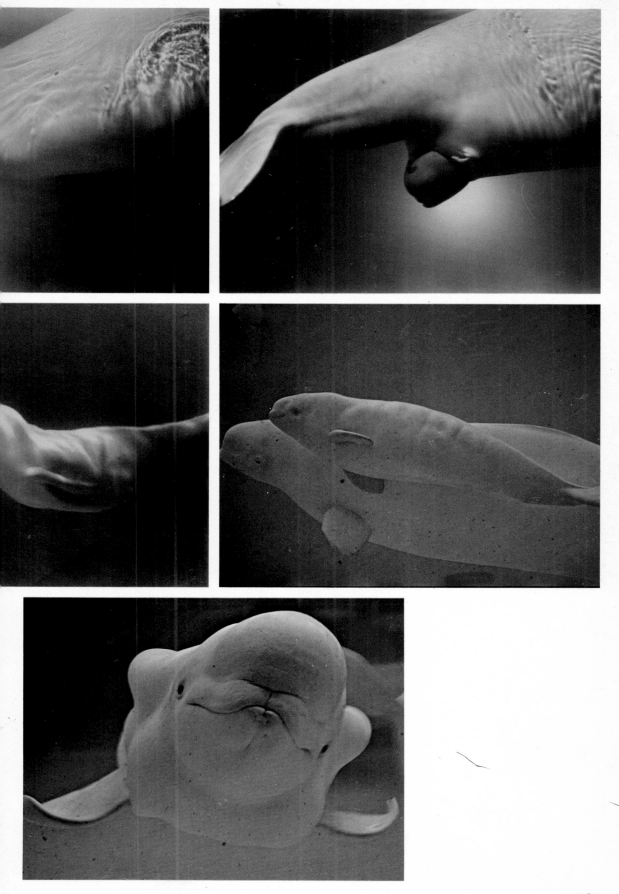

whales. Another major event is attainment of maturity. In sperm whales, for example, both male and female become sexually mature at about 9 years of age, although they don't reach full growth for another 25 or so years. Among the males, usually only the largest, strongest bulls are sexually active, breeding with as many mature females as are in the family pod.

The act of coitus among large whales has not been observed in detail, although whalers and scientists agree there is a great deal of splashing and thrashing about, perhaps in foreplay. It is believed their sex act is completed with both animals vertical in the water, with their heads exposed and facing each other. Coitus among dolphins in tanks, however, has been closely observed and photographed on a number of occasions, and intromission is rapid. Perhaps the same is true for whales.

The human gestation period is 9 months; a pregnant whale cow carries her young for a term of 16 months. Early in its development, the fetal whale only slightly resembles its giant parents. A whale fetus just 10 centimeters long, examined aboard a whaling ship, had a head that resembled a pig, with two distinct nostrils on the snout. During its 16 months within the mother's body, the fetus changes shape and grows greatly in size. One nostril closes over and disappears, while the other shifts to the top of the head and becomes the blowhole, or breathing apparatus. Finally, after nearly a year and a half in the womb, the whale calf is born. Unlike most other mammals, which usually are born headfirst, the whole calf is born in a caudal, or tailfirst, presentation. If it were born headfirst, as sometimes happens, the young whale could drown during birth.

The birth process is rapid: the calf slips easily through the birth canal. As soon as the calf is free, the mother twists her body violently to break the umbilical cord. She quickly escorts her calf to the surface to breathe deeply for the first time in the air. The calf weighs about 1 ton at birth and is nearly 4 meters long. For a time the mother must support him near the surface with the huge bulk of her body; but soon the calf learns to swim and breathe on its own. Very quickly it begins to search for the nipples that will sustain it with milk for nearly the first 2 years of its life, milk that is nearly 34 percent butterfat. (By comparison, milk from a holstein cow averages about 5 percent butterfat.) The rich milk enables the calf to gain weight at the rate of about 3.5 kilograms per day.

By the time the young calf is 2 years old, it has begun to feed on fish and squid, supplemented with its mother's milk. But the female grows increasingly irritated by the calf—no longer small—and its frequent bumping and rushing at her breasts. Soon she forces it away and begins a resting period of about 8 months before again becoming receptive toward the leader of the pod. This cycle will continue for another half century until death intervenes, rarely from extreme old age (say, 75 years) but usually at the hands of a harpooner aboard a killer whaling ship.

The Eternal Wanderer
Far overhead, above the southern ocean, the wandering albatross (*Diomedea exulans*) soars on its 4-meter wing-

The prenuptial displays of albatrosses (Diomedea exulans) *are indeed complex. Prancing about with fanned tails and outstretched wings, the birds touch bills and bow elaborately to each other. For a full year the adults will care for their young, flying as far as 4,000 kilometers from the nest but returning periodically to feed the chicks.*

span, seemingly detached from all that is going on below. The largest flying marine creature on earth, the albatross is one of the few birds that have learned to live in remote ocean waters. Effortlessly riding on the thermal air currents caused by the sun's heat, it may fly halfway around the world in a month of wandering.

Turning at the summit of a climb, it plunges down toward the ocean at a speed of 65 kilometers per hour. By the time it reaches the sea, its wings slicing through the waves, its speed will have accelerated to 145 kilometers per hour. The wandering albatross sleeps on the surface of the restless sea, drinks seawater, and feeds on cuttlefish (*Sepia*). Albatross visit the shore only to breed. A single chick is produced, and for one full year the parents must care for the offspring before it fledges. Although they return to shore for feedings, in the intervening periods the adults may fly more than 4,000 kilometers from their nest.

Exploitation of the Open Sea

The surface waters of the open sea are very different from the shoreward parts of the world ocean. They are populated by an immense variety of organisms—some commonplace and some bizarre—that are never seen by landsmen unless these organisms accidentally drift ashore. But this is rapidly changing. Wide-ranging research vessels have revealed the richness of the open seas of the world, particularly the fishes suitable as food for burgeoning human populations. Research vessels have quickly been followed by large and extremely efficient fishing vessels, capable of catching hundreds of tons of fish in a single day. Unfortunately there is no effective worldwide control over these vessels, and there are few long-term international plans for managing the sea's resources. While some vessels are busily and efficiently removing materials from the sea, other vessels are just as actively dumping materials *into* the sea. Much of the dumped material includes substances that are extremely harmful to marine organisms. The roster of dumpings include sanitary wastes, solid wastes, and petroleum products. Although oil spills offshore garner the most publicity, the bulk of petroleum in the marine environment is from shoreside depots or from tankards. Fortunately, much of the petroleum is eventually degraded by the sun and the air and by marine bacteria that consumes the oil as food. Of far greater concern is the dumping of sanitary wastes and heavy metals, including mercury, cadmium, and lead into the sea. Bacteria and viruses from human wastes have infected fishes that swim through dump sites. And the metals are absorbed in the tissue of fishes rendering them unfit, indeed poisonous, in the human diet. Responsible groups in the world community recognize that the dumping of wastes into the sea is a global problem. Many individuals and institutions are working toward the goal of controlling the dumping for effective management of all the resources of the world ocean. Without such controls and management, those parts of the world ocean may well become the biological "deserts" they once were thought to be.

164–165. *The wandering albatross, the largest albatross, averages 9 kilograms, with a wingspan of 3 meters. Its large wings allow it to ride air currents for long distances while making few wing flaps. Individual birds, identified through marks painted by observers or peculiarities of plumage, have reportedly followed ships at sea for as long as 6 days. Banding studies have revealed the wandering albatross may cover tens of thousands of kilometers each year.*

The Great Depths: Descent into the Abyss

If observers once believed the open sea was virtually a biological desert, then the dark abysses of the sea must have seemed even more desolate. Even the foremost scientists of the day thought so. The bottom of the sea was thought to be a featureless plain that stretched for thousands of kilometers from shore to shore. Research voyages and, later, the records from sonar fathometers, however, showed the sea floor was far from unsculptured and had even more spectacular landscapes than found on dry land. Some seabed mountain ranges are taller than the Himalayas, and the towering peaks of the great mid-ocean ridges soar tens of thousands of meters above the bottom. Here and there in the seabed, deep rifts like the Puerto Rico Trench and the Marianas Trench plunge precipitously; one could tuck the lofty peak of Mount Everest in such trenches and still have nearly a kilometer of water between the mountaintop and the surface of the sea.

But it was believed that even if there were mountains and valleys on the ocean floor, surely the crushing pressure of the water on the bottom, the absence of light, and the probable lack of food excluded all possibility of life in the abyss. But painstaking research by scientists from many nations has revealed that pressure itself is no hindrance to life in the depths of the sea. The deep-sea creatures are able to compensate by developing an inner pressure to balance the outer pressure. Studies of whales, walrus, and seals have shown they have no difficulty making forays from the surface into the depths. Harpooned whales were found to plunge to depths of 800 meters, and it was reasoned that sperm whales must descend to great depths in their normal feeding excursions, since the giant squid on which they feed are creatures of the lightless abyss.

The absence of sunlight is no problem, it was discovered, because most of the inhabitants of the zone of perpetual midnight furnish their own light in a variety of ingenious ways. And while food is not plentiful, it is there to be had even though some curious modes of feeding are necessary.

Realm Without Life

As scientists increased their systematic studies of the world ocean, they began to gain understanding of the zonation of life in the sea. The infant science of ecology—the study of living things in relation to the environment—had not yet been given that name, but many of its principles, based on studies on land, were well known. In order to live, organisms must have light for plants to grow, food for subsistence, and adequate temperatures in which to carry on life processes. Thus, well into the nineteenth century, most scientists agreed that no animals could possibly survive in the numbing cold, eternal night, and overwhelming pressures of the abyss. But this firmly established concept was disproved as the result of another of man's oceanic endeavors.

In 1858, the first telegraph cable was laid on the Atlantic Ocean floor to link Europe and North America. In its day, this engineering feat was as technically complex and aroused the same public interest as moon explorations do now. At last, statesmen and businessmen could converse within a matter of minutes or hours across the storm-tossed Atlantic rather than wait days or weeks for messages carried on ships. The early transatlantic cables failed frequently, however—broken by some accident hidden beneath kilometers of water. To repair them, the cables had to be caught with many-pronged grapples and carefully raised to the surface.

As the shiny black cables, still dripping with seawater, were stretched out on the deck of a repair ship, the workers gazed in astonishment. The cables' surface was here and there covered with strange growths never seen before. Closer examination showed these growths to be living organisms.

The organisms encrusting the cables demonstrated that life was possible on the sea floor, at least to the 2,000-meter depth at which the cables rested. But not all scientists were convinced. Many still clung to the idea of an azooic zone, a barren depth in which no organisms could survive. Other scientists, however, were inspired by the finds on the cables to probe deeper and deeper into the sea in search of life.

Thus, Wyville Thompson, a young English biologist, dredged the bottom at a depth of nearly 4,500 meters and brought up living organisms with each haul. The depth at which such hauls could be made was limited only by the length and strength of the line attached to the dredge and the ruggedness of the equipment. Thermometers attached to some dredges confirmed the icy chill of the great depths.

A dramatic illustration of the low temperatures and extreme pressures of the abyssal zone was provided by Alexander Agassiz, son of Louis Agassiz, the Swiss-American pioneer in oceanography and marine biology, during a research cruise of the *Blake* late in the nineteenth century. The younger Agassiz had been sorting specimens brought up by a dredge in tropical waters. The day was hot and steamy, but the specimens were cold. Agassiz decided to take advantage of the cold depths to take some refreshment. He attached a bottle of champagne to the dredge and sent it to the bottom, 4,300 meters down, for one hour. Returned to the surface, the bottle and its contents were well chilled but undrinkable. At that depth, the weight of a water column exerts a pressure of nearly 120 kilograms per square centimeter. The cork had been driven deep into the bottle, so that the champagne was forced out and replaced by cold seawater!

For many years, the organisms collected in the ocean depths were mostly invertebrates, with only an occasional fish. However, a record was set in 1901 when Prince Albert of Monaco, who made many valuable contributions to marine science, caught a fish, later named *Grimaldichthys profundissimus*, at 6,000 meters off the Cape Verde Islands. For half a century, this specimen held the record as the deepest-dwelling fish ever brought to the surface.

Adaptations to the Depths

The collections of fishes and invertebrates from the great depths had a strong influence on biological science. Not only were new, often bizarre animals recovered from the abyss,

168–169. Despite its name, the viperfish (Chauliodus) is a small denizen of the deep ocean that reaches a maximum size of 20 centimeters. It is known mainly from net captures between 400 and 2,000 meters. During the day viperfishes are found at great depths, but at night they migrate upward, where food organisms are more plentiful. Some species have light organs in their mouths, with which they lure their prey.

The scaleless dragonfish (Bathophilus metalicus) is known only from specimens netted at depths of 200 meters or more. It reaches a length of about 15 centimeters and feeds on small crustaceans. Its earstones (otoliths) have annual growth rings that suggest a life-span of 8 years.

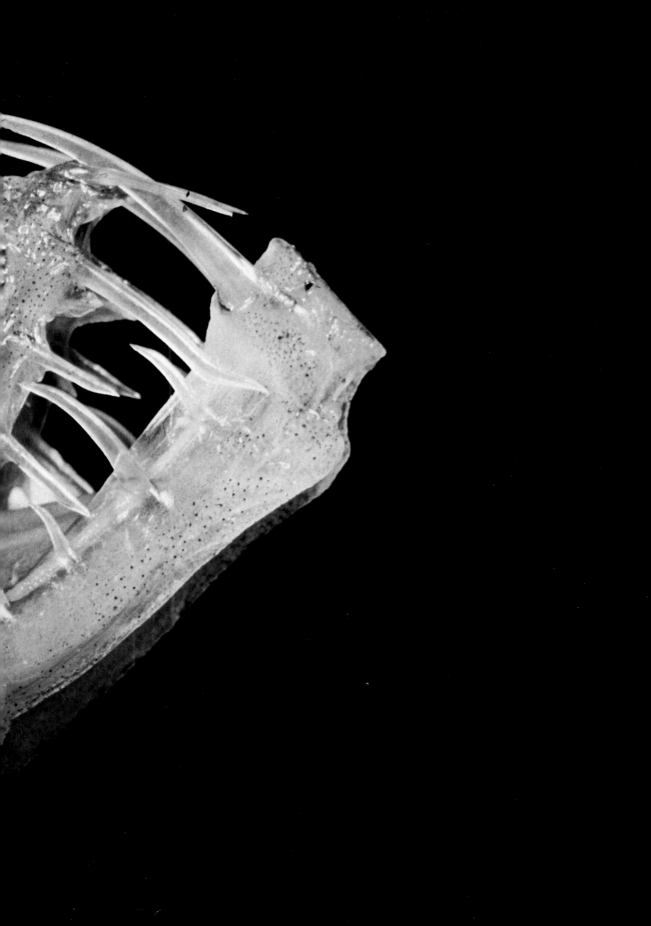

but they exhibited fantastic adaptations to the cold and dark of the ocean depths. They also led scientists to speculate about what greater discoveries might be made— what monsters, perhaps even fabled sea serpents, might be lurking just a little bit deeper.

Biologists noted that one of the major adaptations to life in the great depths was coloration. Fishes and invertebrates on or near the surface were brilliantly arrayed in blues, greens, and yellows; in contrast, animals collected in the abyss were mostly dark red or simply black. Their skeletons and shells were weak, and some had gelatinous tissue replacing muscle tissue. The differences in skeletons and muscles, it was found, resulted from the absence of strong water movement for the animals on the bottom to struggle against. The absence of light was no problem: the animals either were blind or had overly large eyes to collect the maximum amount of light that flashed from the light-emitting organs which seemed present on every creature.

The abyssal fishes were mostly nightmarish animals with jaws that opened wide to reveal cavernous mouths armed with cruel teeth. In some fishes, the mouth led to an extensible stomach that enabled the animal to swallow very large prey. Because prey species were few and scattered, the predators were fast-moving, rapacious attackers. Curiously, below about 10,000 meters, there were almost no predators, and most of the organisms found were bottom-deposit feeders.

Depth Zones in the Sea

The world ocean has been divided vertically into two large divisions: pelagic zones, that is, zones where plants and animals live in the water independently of the bottom; and benthic zones, where the organisms live on or very close to the bottom. The two zones are further sub-divided, according to depth. In the pelagic zone we find the surface dwellers (epipelagic), the middle dwellers (mesopelagic), the deep-sea dwellers (bathypelagic), and the deepest, or abyssal, dwellers (abyssopelagic or hadopelagic).

Some of the benthic zones—for example—the shallows zone, have been discussed in earlier chapters. Others include the bathybenthic, the abyssobenthic, and the most profound depths of all, the hadobenthic. Each of the overall zones—pelagic and benthic—contains animals as different as the zones themselves.

The epipelagic zone encompasses the sunlit waters of the sea we are most familiar with. Here are found tunas, herrings, flying fish, dolphins, and whales—the greatest quantity of life in the open seas. In clear, subtropical waters this zone may extend downward 300 meters.

The mesopelagic, or twilight, zone lying below the lighted zone is between 700 and 1,000 meters down and is home to the tiny, thin hatchetfishes (*Argyropelecus*, *Sternoptyx*), whose common name aptly describes their shapes. Here, too, are the lantern fishes (*Diaphus*, *Myctophum*), darting about and lighting the twilight zone with brief flashes of pale light from their photophores. This is also the vertical realm of the most nightmarish-looking creatures, the

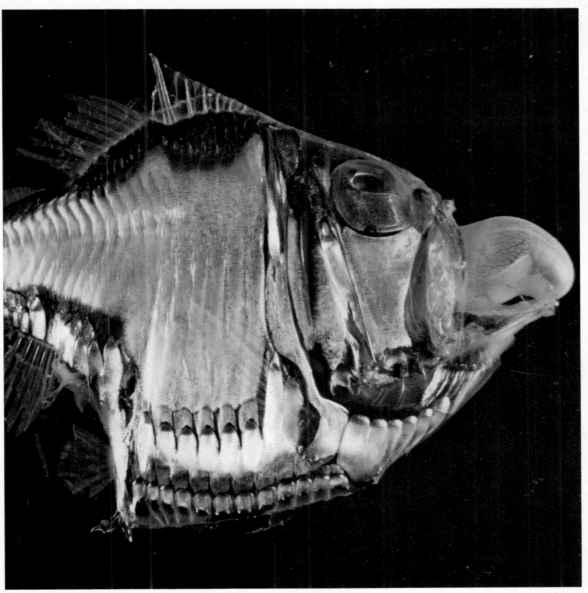

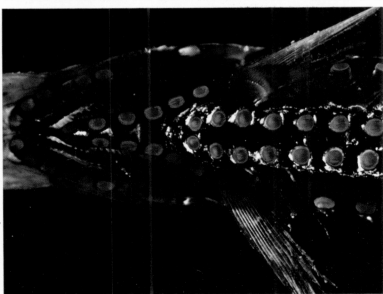

Above. *Resplendent with its
many light organs, the deep-sea
hatchetfish (Argyropelecus)
seldom exceeds 8 centimeters in
length. Fishes brought up from the
depths often suffer a swollen swim
bladder, which pushes the
stomach out through the mouth.
Left. A ventral view shows a
hatchetfish's light organs,
arranged in rows.*

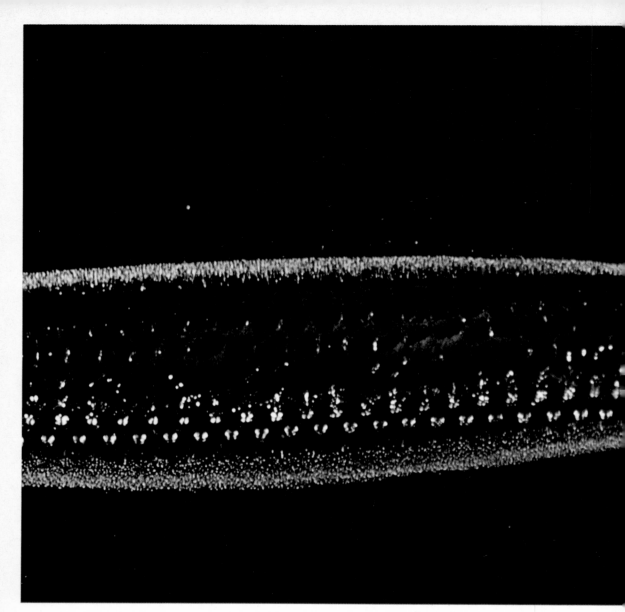

Scaly dragonfishes (Stomias) are elongate and reach about 15 centimeters in length. Nearly all have some sort of whisker (barbel) attached to the chin. The barbel may be five times as long as the fish itself and may be equipped with one or many light organs. Other light organs occur in rows along the body. Dragonfishes have toothy jaws that cannot be tightly closed.

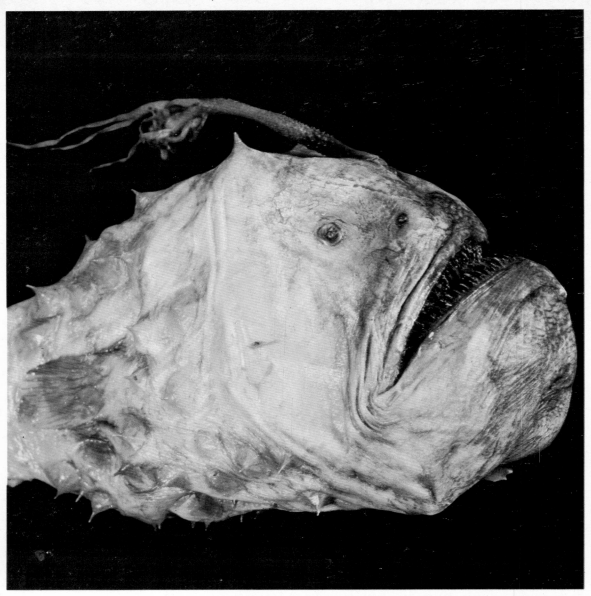

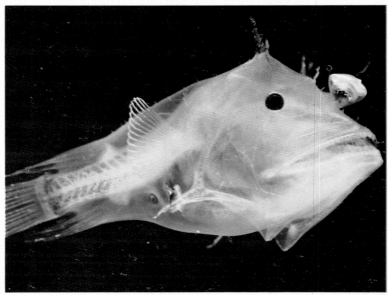

*Deep-sea anglerfishes were first discovered by a Danish naval officer, who found specimens washed ashore on the west coast of Greenland in the early 1800s. Since then, many more specimens have been collected from deep water in all oceans. Of about 80 species currently recognized, most are only a few centimeters long, but some reach a meter or more in length. Deep-sea anglers are remarkable for the habit of male parasitism. The males are small, about 1 centimeter in length, and early in life become permanently attached to a female and appear as mere fleshy appendages of her body. Attachment is accomplished by the male's biting the female and holding on. Ultimately, as the attachment becomes permanent, the digestive system of the male degenerates and the male and female tissues fuse together. Female anglers accumulate males throughout their life. Being dependent on the female, the attached males die with the female. Top. A large (about 400-millimeter) female of the first species ever described (*Himantolophus groenlandicus*). Bottom. A specimen of another, unidentified species.*

viperfishes (*Chauliodus*) and the dragonfishes (*Stomia*), and of the light-emitting squids (*Spirula, Lycoteuthis, Ommastrephes*).

Entering the eternal midnight of the great depths, we come to the bathypelagic zone, between 1,000 and 2,000 meters, and the abyssopelagic zone, from 2,000 meters to just above the ocean bottom. (Below 6,000 meters, scientists speak of the hadopelagic region.) These uttermost depths of the sea are populated by fishes called gulpers and swallowers, by dark red squids (*Histioteuthis*) with membranous webs between their tentacles, and by a strange blind octopus (*Cirrothauma*) with projections on its body resembling a pair of small wings.

The gulpers, or pelican eels, swim through the inky gloom watching for telltale glimmers of luminescence from a careless squid or even another gulper. Gulpers vary in size from the 60-centimeter *Eurypharynx pelecanoides* to *Saccopharynx ampullaceus*, nearly 2 meters long. Most of a gulper's length is in its slender, whip-like tail, and the rest is its mouth, actually a roomy bag armed with teeth. The gulper's eyes and brain are tiny, but its mouth is huge; the jaws unhinge, and its belly expands to several times its normal size. A gulper can thus swallow a fish as large as itself or even larger. Indeed, Albert C. L. G. Günther, the famous nineteenth-century German-born fish biologist, said that gulpers do not *swallow* their prey but, like a snake ingesting a mouse, ". . . draw themselves over their victim" and store it in a capacious belly.

Finding food in the inky depths is a matter of opportunity, and the predator must be prepared to take what it can when it can, regardless of the size of prey. The problem has been solved in a unique manner by the anglerfishes. The bathypelagic anglers resemble in miniature the fishing frogs, or anglerfishes (*Lophius piscatorius, L. americanus*) of the North Atlantic. But whereas the fishing frogs may be 1.5 meters long, many deep-sea anglers are only 5 to 10 centimeters long. These little fishes, about the size of a man's fist, dangle a lure—like a fisherman's bait—on a small "fishing rod" just in front of a gaping mouth rimmed with saber-like teeth. Prey investigating the lure are quickly engulfed and digested by the angler.

The problem of finding a mate also has been solved in a unique manner by the anglers. The male of the deep-sea angler, very small compared with the female, swims about seeking a mate. When the male encounters a female, he attaches himself to her side with his teeth. Then, undergoing a degenerative metamorphosis, he becomes a small sac containing only the male sex glands. The male now derives his nourishment directly from the female's circulatory system and is ready and available to fertilize her eggs at spawning time. Some females in the bathypelagic zone may each carry 5 to 10 males.

A Direct Look at the Abyss

The mysteries of the sea that were revealed in the contents of each dredge and trawl hauled dripping from the hidden depths served merely to whet the appetite of scientists to see more, especially to see the organisms in their natural habitat. The development of cameras capable of withstanding great pressure, and fitted with lights to illuminate the eternal gloom, soon provided researchers with an opportunity to see exactly what it was like in the sea's abyssal depths.

Underwater television and movie cameras gave tantalizing glimpses of life on the ocean bottom, but technical problems such as lack of clarity and inadequate illumination often resulted in murky viewing. Further, there was no guarantee that any organisms would be in view when a camera was operating. In fact, most undersea viewing at great depths resulted in virtually blank TV screens or hundreds of meters of film recording a seemingly lifeless seascape. Still cameras with high-intensity flashes were little better until an enterprising photographer thought of borrowing a trick from fishermen and "baiting" his camera. A still camera was rigged with a pot of fish scraps placed as bait at a fixed distance from the lens. Lights were aimed at the bait. The entire unit, programmed to flash the lights and take photographs at intervals, was dropped free into the ocean. The unit settled to the bottom, took photographs, and after hours or days, was automatically buoyed to the surface and recovered.

When the films were processed, the results surpassed even the fondest hopes of scientists interested in abyssal creatures. Crystal-clear photographs showed the abyssal floor, its soft mud crisscrossed with curious trails made by unknown crawlers, and small burrows and mounds constructed by creatures beyond the camera's view. But the animals clustered around the bait pot were seen now as whole, living organisms instead of the mashed and torn blobs brought up by the collecting dredge or trawl.

A camera set on the bottom, 1,400 meters deep in the San Diego Trough off southern California, photographed multitudes of grenadiers (*Coryphaenoides*) and sablefish (*Anoplopoma fimbria*) swarming over the bait pot and trying to get at the fish scraps inside. Slime-coated hagfishes (*Myxine*), blind primitive fishes related to lamprey eels, writhed in incredible knots, seeking to fasten their jawless mouths on the bait. Surprisingly, the photographs also revealed very large sleeper sharks (*Somniosus pacificus*) attracted to the bait.

Cameras dropped at various places in the Pacific Ocean at depths of 3,500 to 6,000 meters took photographs of many grenadiers; this frequence suggests they may be among the most common fish species in the great depths. Because of its shape, the grenadier is sometimes called the rattail or marlinspike. It has a large, heavily armored head with a barbel, a "little beard," under its lower jaw. The body tapers rapidly to a long slender tail. The eyes are large and apparently very responsive to what little light may flash from other bottom dwellers. The grenadier's barbel— a very sensitive organ in most fishes, including the cod, a grenadier relative—is delicately drawn over the abyssal floor in a search for food. When a burrowing amphipod or other animal is detected, the grenadier digs it out of the bottom, like a pig rooting out a succulent tuber, and sucks it into its downward-directed mouth.

Grenadiers vary in size, depending on the species and where it is found. Most are 30 to 90 centimeters long. The common grenadier (*Macrourus bairdii*), from the North Atlantic and around the Azores, is one of the smaller species. The rough-headed grenadier (*M. berglax*), from around Iceland, northern Norway, and southern Greenland, is one of the larger species. But all grenadiers are uniformly drab-looking, ash gray on both top and bottom. Some specimens of the common grenadier caught in trawls towed at abyssal depths are tinged with color—pinkish snouts, violet throats, pink dorsal fins, and dark blue eyes.

The undersea cameras revealed the true shapes and appearance of other deep-dwelling fishes, including the brotulids, as represented by *Tauredophidium hextii* and *Leucicorus lusciosus*. (These fishes are so rare they have only the Latin names given them by ichthyologists and taxonomists.) In general, the brotulids resemble the grenadiers, except for their eyes. Brotulid eyes are small and hardly function at all as organs of sight. But these fishes more than compensate for this lack of sight—not really a lack in the perpetual blackness of the abyss—with a very sensitive lateral-line system. With this system, a brotulid can detect the faint water currents set up by a squid or another fish swimming nearby. A truly blind brotulid (*Typhlonus nasus*), collected in trawls dropped to 3,140 meters in the Celebes Sea, south of the Philippine Islands, depends entirely on sensing vibrations with its lateral line to detect prey and avoid possible predators. Like so many other strange fishes of the great deeps, the brotulids are rather small. *Typhlonus*, for example, is only about 22 centimeters long.

"Mousetrap" and "Tripod" Fish

The abyss is home to creatures that defy the imagination with their strange shapes and striking adaptations to existence in this zone of eternal night. Two species in particular look like visitors from outer space. One of these, *Galatheathauma axeli*, was collected by the Danish *Galathea* expedition from a depth of 3,590 meters off tropical Central America. A type of anglerfish, it was nicknamed "the living mousetrap"—with very good reason. A large, lighted and forked lure hangs from the roof of the mouth behind pointed, curved teeth that rim the jaws like the teeth of a comb. The fish, 47 centimeters long, undoubtedly rests on the bottom, its mouth agape, displaying this glowing lure in the stygian gloom. When a luckless fish or prawn comes close to investigate the "bait," the great toothed jaws snap shut around the prey.

The other "outer space" organism is a small fish about 25 centimeters long with thin, flexible extensions, each about 35 centimeters long, from the lower portions of its caudal (tail) and pelvic fins. These extensions are actually elongated fin rays, and when scientists saw the first specimens that fell out of a trawl, they called it the "tripod fish." At first they thought the tripod was used simply to detect food organisms buried in the bottom. Later, viewers in the bathyscaphe *Trieste*, diving in 7,000

178–179. Crustaceans are probably more numerous than any other macroscopic life forms in the ocean. Shrimp-like creatures, such as this euphausiid, are particularly abundant in some areas. In surface waters they constitute the krill on which the large baleen whales feed.

Below the surface and at mid-depths, the sea is only sparsely populated with fishes. When one fish meets another, one is liable to swallow the other forthwith. Large size is no advantage in the deep ocean. What counts is a big mouth and an accommodating stomach. Quite possibly, the swallower will be a small fish with a big stomach. This specimen, probably a hammerjaw (Omosudis lowii), is only about 100 millimeters long but gives every indication of having swallowed another fish two or three times as large as itself.

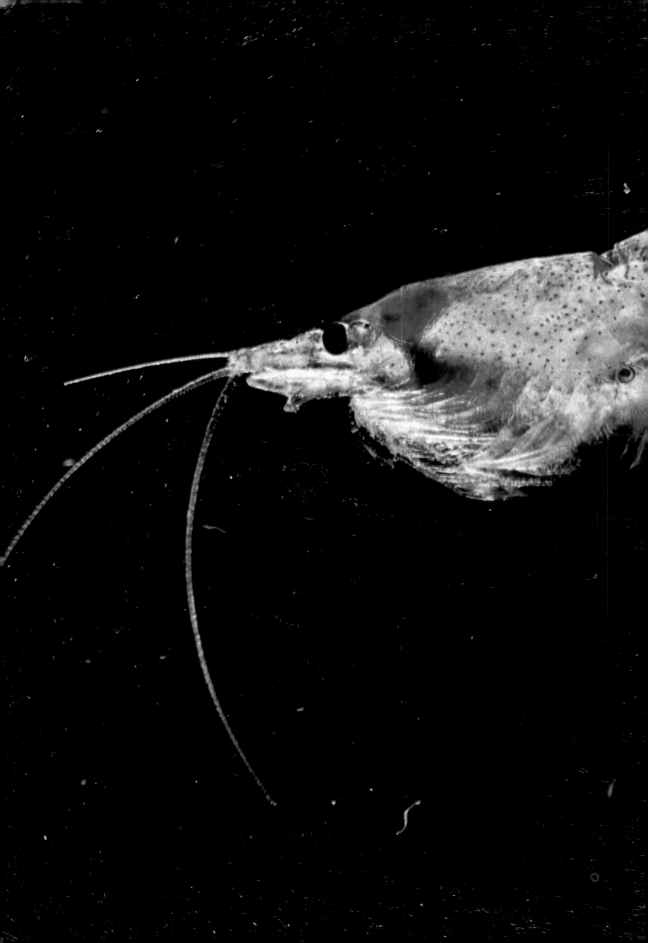

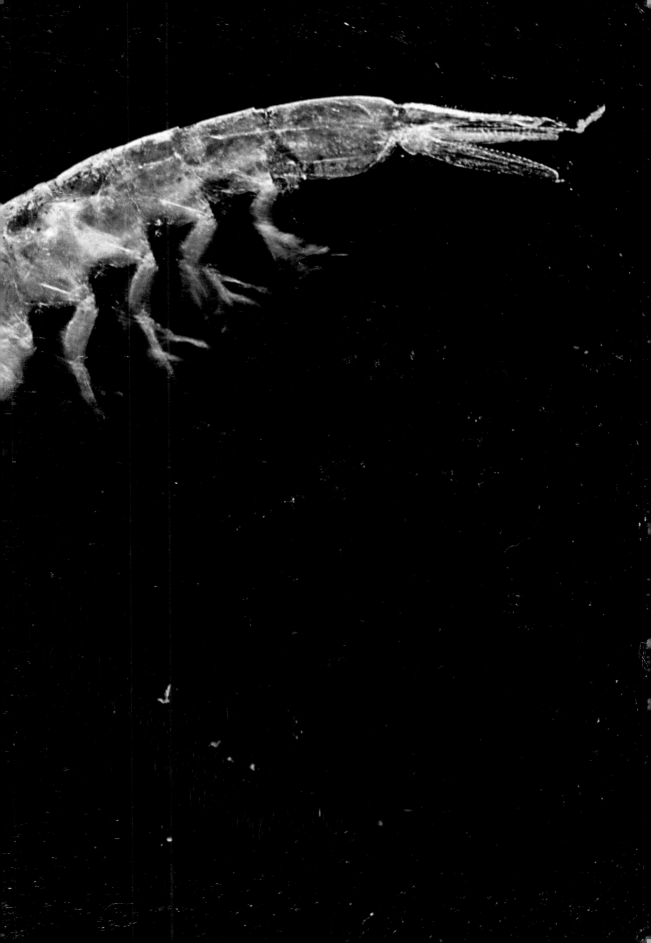

meters off the Pacific island of Guam, watched with wonder as a tripod fish (now named *Bathypterois viridensis*) used its three fin rays, like a tripod, to stand on the bottom. Another, more revealing observation of a tripod fish was made from *Deepstar 4000*, a research submersible designed by French oceanographer Jacques-Yves Cousteau. The vessel was hovering above the bottom at a depth of 1,219 meters in the Gulf of Mexico off Florida. The scientists saw a tripod fish about 25 centimeters long perched majestically on its pelvic- and caudal-fin tripod, with sail-like pectoral fins arched above its body like the arms of a ballet dancer. When prey came within range, the fish leaped from its three-point stance to snatch at the morsel.

Invertebrates of the Abyss

The sea cucumbers, sea urchins, starfish, and crabs of the great depths have had to make many adaptations to the hostile environment of cold, pressure, and darkness. Most of these abyssal organisms bear only a family resemblance to their relatives of the shallows zone. When the *Galathea* completed a 7-hour dredge haul—from the surface, on the bottom, and back to the surface—to a depth of 4,820 meters between Madagascar and Mombasa, scientists found that the most abundant animals retrieved were sea cucumbers. (These soft, slimy, cylindrical animals are paradoxically included with the starfish and sea urchins in the group called echinoderms, "spiny-skinned.") Two species of sea cucumbers were *Deima*, whitish and about 10 to 15 centimeters long, and *Psychropotes*, reddish-violet with a yellowish back and about 20 to 30 centimeters long. *Psychropotes* has a semicylindrical body and a great tail appendage rising like a sail behind it. The "sail" may be used weakly as a swimming organ. Scientists believe the animal ploughs its way through surface mud, holding its tail well above the bottom as a respiratory organ while gorging itself on the deep-sea ooze. Similarly, sea urchins in the abyss bear little resemblance to the creeping pincushions of the shallows and the rocky shores. The very rare *Echinosigra paradoxa*, taken by the *Galathea* at about 3,000 to 4,000 meters, does not have the typical oval shape of other urchins. Its rear portion is drawn out so that the entire creature looks like a head on a long neck creeping across the bottom ooze. This species has an unusual distribution: it was collected at about 1,515 meters off southern Iceland and at about 3,300 meters in the middle of the Indian Ocean.

Food on the Bottom

The deep-sea cucumbers and urchins live a precarious existence, depending on bits of organic matter in the bottom ooze for nutrients. They must contend with numbing temperatures, ranging from −2°C in Antarctic waters to 1.5°C in the depths of the Pacific, and with pressures as great as 400 to 700 kilograms per square centimeter. But the source of their food is the most vexing question. Undoubtedly some nutrition comes from the dead and decaying bodies of other dwellers of the midnight depths; and some, though not very much, must come from surface organisms that die and drift slowly downward. But

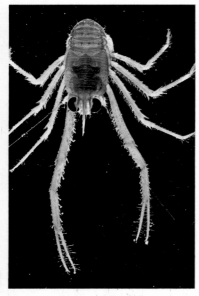

Crustaceans are ever-present in the ocean. This deep-sea anomuran is thought to be related to hermit crabs, king crabs, and lobster krill.

recent analyses and laboratory studies do not bear out the picture presented by popular writers of a perpetual "snowfall" of organic matter filtering down through thousands of meters of water. In reality, very little of the surface organic matter ever reaches the bottom. Most animals that sicken or die at the surface are eaten as they drift downward. Microscopic copepods sink at the rate of 500 meters per day, and thus would take a week or more to reach bottom in deep water. As they slowly fall, they are snatched up by fishes, jellyfishes, and perhaps other copepods. It has been estimated that a dead whale sinks at the rate of 100 meters per hour and takes about two days to reach bottom in the abyss. As it sank, parts of its carcass would be savaged by sharks and even by squids. Because they are consumed as they fall to the depths, whale carcasses probably contribute less than a gram of food per square meter per year for the abyssal bottom feeders. Even the bones of whales and larger fishes are rapidly consumed, leaving only the hardest tissue untouched. The most common solid animal remains on the bottom are waxy earbones of whales and sharks' teeth. One dredge haul on the bottom of the Pacific Ocean at 4,300 meters yielded 50 whale earbones and thousands of sharks' teeth.

Venus's-Flower-Basket

Sponges are simple, humble dwellers in the depths of the sea. These colonial animals—or at least their dried and softened bodies—serve us for scrubbing in the bathtub and washing the family automobile. But one group of deepwater creatures, the glass sponges, are as dangerous to touch as they are beautiful to look at. As their name implies, these sponges have a "skeleton" of pure silicon, clear as crystal and just as brittle. In collecting trawls, sponge fragments can cut a crew's fingers as nastily as splinters of glass. The main body of the sponge grows up from the bottom on a strong rope-like stalk of twisted silica strands. Deep-sea barnacles, *Scalpellum*, and sea anemones take advantage of the stalks and attach themselves to keep above the bottom ooze.

The silicon spicules of the skeleton are fused to form a lattice. When the skeletons are dried, they look strikingly like flower vases of woven strands of glass. Many people keep them as curios, and little wonder they are so poetically called Venus's-flower-basket. Some species of glass sponge (*Euplectella*) have an interesting association with abyssal shrimp: young male and young female shrimp may swim into the hollow core of the sponge and remain there until they grow too large to swim out. Feeding on organic matter that enters with the sponge's respiration, they spend their entire lives in this organic spongy prison. In days gone by, sponges with imprisoned shrimp were given as wedding gifts in Japan, symbols of the wedding vow, "Till death us do part."

Deep-sea Crabs and Squids

Crabs are versatile animals, well able to exist and even thrive in the most difficult marine environment. The conditions of the abyss are no exception. In water as moderately deep as, say, 1,500 meters, some species are quite common. One in particular, the red crab (*Geryon quinquedens*), remained undetected for many years simply because no one was fishing or collecting in areas inhabited by it. The species was little known until well into the 1950s, when some experimental deepwater trawling by a research vessel revealed its existence. The crab was considered somewhat of a curiosity because of its bright red shell and because it sometimes weighed a kilogram or more. The red crab became better known when many were caught in lobster pots in deep water off the U.S. east coast. Soon after, single vessels were bringing in as much as 635 tons of the crabs each year for the food market. Thus, in about twenty-five years, the red crab changed from a little-known deep-sea curiosity to a prized food item. Exposed to such unrestricted fishing, it is possible the crab may be overfished and truly become a rare curiosity once more.

There is little danger, however, that other abyssal crabs will become overfished. One reason is that they are much too few in numbers, and dredging for them in depths of 2,000 meters or more is beyond the capability of any but specially equipped research vessels. In addition, these crabs do not yield much in the way of edible flesh. The deepwater crabs *Ethusa* measure only 15 to 25 centimeters across, from claw tip to claw tip. Like most dwellers of the perpetual night of the abyss, they are blind. Their shell is a pallid hue, although the eggs the females carry under their tail may be yellow-orange.

Even hermit crabs (*Parapagurus*) scurry about on the floor of the abyss, much like their counterparts in the sunlit shallows zone hundreds or thousands of meters above. Here in the great depths the hermit crab busily searches for a new shell to live in and for bits of food to scavenge. Because of the scarcity in deep water of shelled gastropods, snail-like animals, the hermit crab's search is an unending, quite frantic one.

Scientists believe the abyss is inhabited by many squids, but few have been collected from such depths because they are agile swimmers and can ecape the trawls sent down to catch them. The abyssal squids that have been collected are as soft and fragile as jellyfish: unlike the species of the shallows zone, they have gelatinous rather than tough muscle tissue. Squids have been collected at depths of 3,500 meters, and in the Wedell Sea off Antarctica an octopus (*Grimpoteuthis*) lives on the bottom at 3,000 meters.

Most squids and octopus in the shallows zone are more or less pinkish, but can change color to match their background. The abyssal cephalopods are different, for some are red, others are black or deep purple, and yet others are translucent or appear almost transparent. The squids sometimes resemble underwater neon signs because they have numerous, multicolored photophores, or light organs, which flash in the gloom as if, to quote one observer, "some unseen artist was playing a musical instrument." At night the squids frequently make long excursions upward into the mid-depths to prey on hatchetfishes and shrimps. With the coming of daylight,

Squids, like snails and clams, are mollusks. Closely related to octopuses, squids have a well-developed brain and a pair of eyes much like those of humans. Whereas octopuses have 8 arms, squids have 10—two of them a little longer than the others. Squids are strong, fast swimmers that develop jet propulsion by pumping water in and out of the mantle cavity, a compartment containing two feathery gills in the animal's underside. The jet can be directed either forward or backward by movement of the funnel-like opening. Giant nerves control the muscles that pump the water and enable the animal to respond to stimuli very quickly. Because of their fast reaction, few deepwater squids are caught in nets. Top to bottom. *Three deepwater species of squids are:* Phasmatopsis; Abralia; Calliteuthis.

183

Many of nature's life forms are round, such as, top, Gigantocypris, *the 1-inch "giant" among ostracod crustaceans and,* center and bottom, *two deepwater jellyfishes. Ostracods are among the most abundant crustaceans, populating virtually all bodies of water the world over. Among crustaceans, they are remarkable for their small number of appendages and segments and for their enclosure in a bivalve shell often mistaken for that of a mollusk. Jellyfishes resemble upside-down sea anemones or coral polyps. Many species of marine jellyfishes include a polyp phase in their life cycle.*

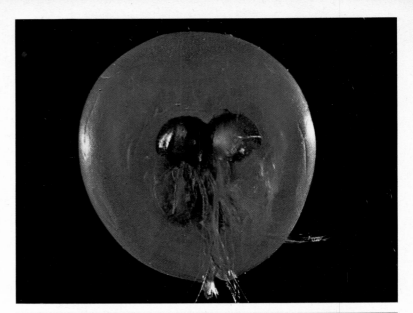

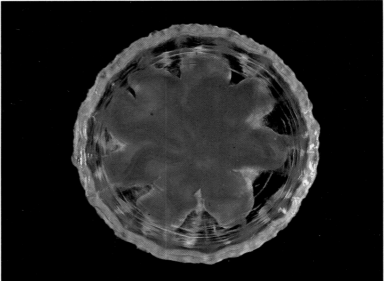

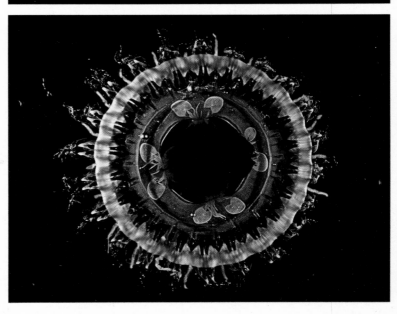

like creatures in a horror story, they descend again into the constant midnight of the abyss.

The Origin of Life and Living Fosils

The new and startling discoveries made in the depths of the sea in the nineteenth century led many scientists to believe that perhaps a primordial creature—the type of primitive animal from which all life on earth arose—might someday be dredged from the depths of the abyss. Many strange and unusual animals did indeed tumble from the deep-sea trawls and dredges, but all of them resembled, in some fashion, animals already known. Then, in 1857, in samples of sediment from the Atlantic floor, a curious gray gelatinous material was noted under the microscope. The material seemed to have many qualities that fit the scientists' idea of what our ancestral life should be like. Excitedly, biologists studied the find and finally announced that the origin of life had been discovered. This amorphous "creature" was even given a scientific name: *Bathybius haeckelii*. But a chemist soon pricked the bubble of hypothesis. To the chagrin of the discoverers, he announced that this was no living creature but only a precipitate of calcium sulphate that formed when alcohol was added to seawater to preserve specimens from the deep.

The error of *Bathybius* did nothing to discourage marine scientists, and their dredges went deeper and deeper as new and better equipment became available. Finally, they were able to dredge below 6,000 meters in the Hadal zone, so aptly described as a place of ". . . utter and perpetual darkness, icy cold, crushing pressure—it qualifies as a type of Hades." Dredging in the uttermost deep was fascinating, because no one knew what might come to the surface in the collecting gear. Indeed, scientists such as Louis Agassiz reasoned that the deep sea might very well be a refuge for prehistoric animals known only from fossilized remains.

Agassiz's hope seemed fulfilled when, in 1864, the Norwegian marine scientist G. O. Sars, dredged a live specimen of the sea lily (*Rhizocrinus lofotensis*) in Norwegian waters at a depth of only 550 meters. Despite their plant-like name, these animals are classed among the echinoderms. Their five-armed "flower" is held erect atop a long stem and serves to trap plankton for food. Remains of this species had previously been uncovered only in fossil beds 160 million years old. Additional, deeper dredging revealed the existence of other "living fossils." Surely, and soon, the depths would yield a living fish of ancient form, and perhaps eventually a marine dinosaur.

In 1938, scientific and popular publications trumpeted the discovery of a living coelacanth, a fish biologists knew only from fossil remains about 60 to 300 million years old. The specimen, caught at a depth of 122 meters by a trawler fishing in the Indian Ocean off East London, Republic of South Africa, was named *Latimeria chalumnae*. It was 1.5 meters long, weighed 55 kilograms, and had blue-gray skin, luminescent eyes with a distinctive blue glow, and a vicious-looking mouth with strong teeth. Stout fins that projected from the body, two on each side,

186–187. *Resembling spiders in appearance, and perhaps related to them, pycnogonids creep about on the ocean floor, probably down to the greatest depths. The legs are so long and the body so small that extensions of the viscera are accommodated within the legs. Perhaps the long legs are adaptations for moving over the soft sediments of the deep-sea floor.*

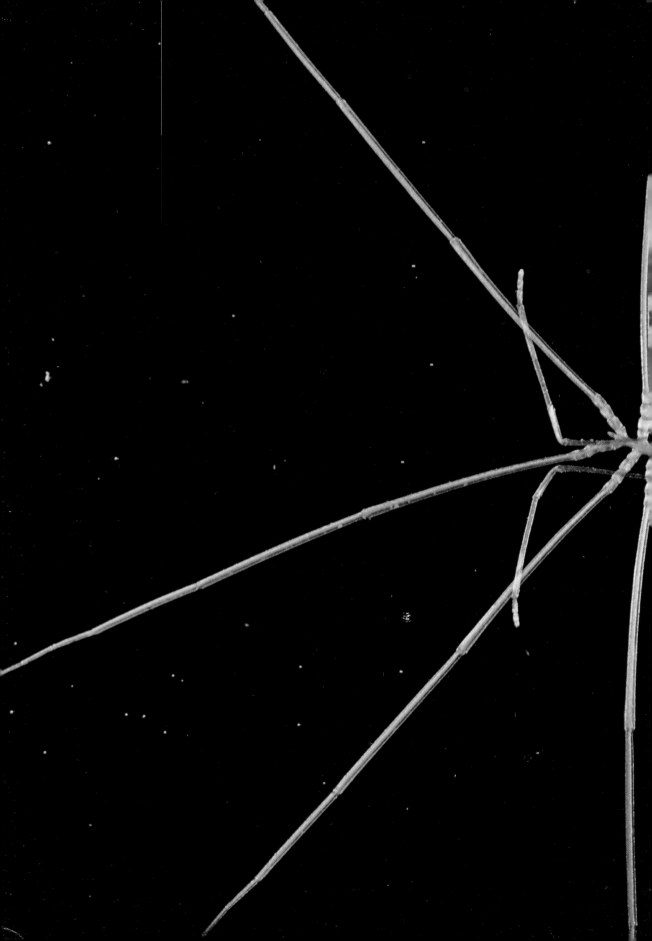

resembled legs and suggested to many that the creature was a missing link between fishes and reptiles. Later, it was found that, in the manner of many sharks, coelacanths bore their young alive from eggs hatched in the mother.

Since the original discovery, additional specimens have been collected, some by native fishermen using baited hooks in depths between 100 and 150 meters. Despite every effort, scientists have been unable to keep coelacanths alive in tanks. The shock of being hauled from the cold of its natural habitat to the subtropical surface apparently is too much for the intriguing fish.

Descent into the Abyss and Return

The abyss is generally characterized as a zone of few living things. But there are patches of abundant life at depths of 2,500 meters in the Pacific Ocean off Central America. Flowing springs of water warmed to 17°C by the earth's molten center create "oases" of marine life. The warm waters and the microscopic organisms they support are surrounded by extensive beds of giant clams (*Vesicomya*), 30 to 40 centimeters across, which grow fat on a rich diet of planktonic life. Large grenadiers (*Coryphaenoides armatus*), about 75 centimeters long, propel their brownish-tan bodies across the clam beds seeking prey. The grenadier's large eyes, among the most sensitive in the animal world, probably help it perceive the flashing, telltale light from bioluminescent prey. The productive ecosystems surrounding the warm abyssal springs are not based on food made by plants as in the shallower photic zone. The abyssal zone lacks the light needed for photosynthesis; instead, the primary source of energy there is hydrogen sulfide gas from the depths of Earth, which nourishes bacteria, the base of the food system. Thus, it is *chemo*synthesis rather than *photo*synthesis that accounts for life in the abyssal oases.

Away from the warm springs, the abyss is, as one might expect, a stable environment with feeble currents where the seasons are imperceptible and there is no night or day. There is little month-to-month variation in temperature, food is scanty, and the organisms exhibit slow metabolism, slow growth, and long lives. Some recent examinations of live clams collected at a depth of 300 meters in the Atlantic Ocean off North America revealed that, although only 3 centimeters long, they were about 250 years old! Occasionally, surface occurrences do affect the abyss. About 55 kilometers northeast of Bermuda, the seabed 5,000 meters down was found to have been scoured by waves generated by hurricanes. But, in general, the abyss is a zone of little change.

Descending into the abyss and returning has always been the dream of scores of men. The discovery of sperm whales tangled in the coils of submarine cables 1,100 meters down was proof that the giant mammals make these forays routinely, mostly without incident. (The stomachs of many were found to contain spider crabs, octopus, and skates—all bottom animals. Perhaps the whales mistook the cables for the massive coils of an octopus or squid.) All diving animals—birds, reptiles, and mammals—exhibit profoundly changed physiology when they plunge substantial distances. The heart rate is depressed dramatically at the beginning of a dive: it may go from a normal 75 beats per minute to 5 beats per minute. Most of the blood, with its life-sustaining oxygen, is shunted to the brain and other sense organs and to the diving muscles, away from the abdominal organs. With these physiological adjustments, the animal is able to function efficiently in the dive. Human beings exhibit diving physiology, but only to a very limited degree, so they must be supported by elaborate mechanisms in order to descend into the depths.

Modern bathyscaphes have provided virtually unlimited opportunity for abyssal diving, enabling humans to descend into the deepest parts of the world ocean. Naturally, elements of risk are always present. The bathyscaphe *Trieste II*, for example, had been exploring the Cayman Trough, south of Cuba, descending to 6,190 meters without incident; then, in one descent, at a depth of 4,570 meters, it collided with the side of the world's deepest volcano. The bathyscaphe was only slightly damaged and was fortunately able to return to the surface.

Human descents into the abyss have revealed the abundance and nature of the animal life that dwells there, as well as the configurations of the seabed. Recently, observers aboard the French bathyscaphe *Archimède* explored the Puerto Rico Trench at the astounding depth of 9,200 meters and saw with wonder that every square meter of sea bottom showed some sign of life—whether a furrowed trail, a low mound, or even a shrimp or two. But the scientists also saw, to their dismay, that the abyss was very vulnerable to human desecration: scattered among the fascinating creatures of the deep were an empty jam tin, scraps of cardboard, and a beer bottle!

The discoveries in the abyss have raised some questions. The abyss is a special kind of environment, because in its darkness and cold, life and physical processes move very slowly. Thus, while man has considered the great depths as a suitable repository for his wastes, including sewage, toxins, spent nuclear fuel, and outmoded munitions (such as the mustard gas in the Baltic), these harmful materials might remain on the sea bottom virtually unchanged, perhaps for centuries. And, in so doing, it is also possible that such noxious substances could gravely damage or even destroy the very fragile, very ancient life of this undersea realm.

Can man destroy the sea? The answer is simple yet complex. No, we cannot physically destroy the world ocean, but we have the capability to damage it seriously. If we are to preserve, protect, and use the sea wisely, as well as its shores and its fauna, the common heritage of mankind, *we must recognize what we are—that is, simply another species dwelling on Planet Ocean—and we must act accordingly.*

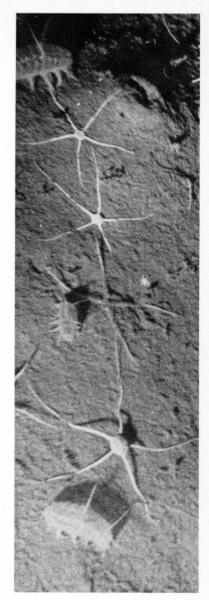

Above. *On the floor of the abyss, as revealed through the observation window of the deepwater submarine* Trieste, *brittle stars, with their five arms outspread, are numerous. So also, marked in the sediment, are the tracks of other organisms. Even in this environment of complete darkness, cold, high pressure, and sparse food supply, most major forms of animal life are present.*
190–191. *At sunset, waves rush toward a Pacific shoreline. Often the source of poetic inspiration, the sea is an enduring reminder to mankind of its evolutionary roots. The sea now represents the human race's hope for survival in a world of rapidly decreasing natural resources.*

Photo Credits

1–9 Maurice Rumboll; 13 Harald Sund; 16 top, second from top Michael Abbey/Tom Stack and Associates, second from bottom, bottom Oxford Scientific Films; 18–19 Jeff Foott; 21 top James David Brandt/Earth Scenes, bottom Breck P. Kent/Earth Scenes; 26 Barry Parker; 31 Harold Simon/Tom Stack and Associates; 32–33 Jerry Cooke/Earth Scenes; 35 Sonja Bullaty and Angelo Lomeo; 39 National Aeronautics and Space Administration; 40 Robert B. Evans/Tom Stack and Associates; 41 Steven C. Wilson/ Entheos; 44 top Leslie F. Conover/Photo Researchers, Inc., bottom Kurt Severin; 47 Richard Chesher/Photo Researchers, Inc.; 48–49 Shelly Grossman/Woodfin Camp and Associates; 51 Colin B. Frith/ Bruce Coleman, Ltd.; 52–53 Row 1: left Jeff Rotman, center Ron Church/Photo Researchers, Inc., right Kenneth R. H. Read/ Tom Stack and Associates; Row 2: left L. M. Stone/Bruce Coleman, Ltd., center L. L. T. Rhodes/ Animals Animals; right K. H. Switak; Row 3: left Alan Power/ Photo Researchers, Inc., center Jeff Foott, right N. Coleman/Tom Stack and Associates; 54 Peter Ward; 55 Jeff Foott; 56, 57 Jack Drafahl; 58 top Calvin Larsen/ Photo Researchers, Inc., center Douglas Faulkner, bottom Verna R. Johnson/Photo Researchers, Inc.; 60–61 Steven C. Wilson/Entheos; 62 top Kenneth E. Lucas/Steinhart Aquarium, bottom Stephen Dalton/Natural History Photographic Agency; 63 top Kenneth E. Lucas/Steinhart Aquarium, center Kjell B. Sandved, bottom Oxford Scientific Films; 64–65 Row 1: left Edward R. Degginger, center Jeff Foott, right Betty Randall; Row 2: left Jeff Rotman, center Klaus Paysan, right Jack Drafahl; Row 3: left Jeff Foott, center Klaus Paysan, right Jack Drafahl; 66 Phil Degginger; 67 top Howard Hall/Tom Stack and Associates; 67 bottom, 68–69 Jack Drafahl; 70, 71 top Francisco Erize; 71 bottom, 72–73 George Holton; 75 Russ Kinne/Photo Researchers, Inc.; 76–77 Jack Dermid; 78 Jane Burton/Bruce Coleman, Ltd.; 79 Carl Roessler; 80 top Charlie Ott/ Photo Researchers, Inc., bottom J. A. L. Cooke/Oxford Scientific Films; 81 top Klaus Paysan, bottom Betty Randall; 82–83 James H. Carmichael; 84 top Klaus Payson, bottom Jack Dermid;

85 Jane Burton/Bruce Coleman, Ltd.; 86 top Douglas Faulkner, bottom Gil Montalverne/Natural Science Photos; 88–89 Jacques Six; 90 Tom McHugh/Photo Researchers, Inc.; 91 Jacques Six; 92–93 Row 1: left and center Hans Dossenback, right Peter Ward; Row 2: left Phyllis Greenberg/ Photo Researchers, Inc., center William R. Curtsinger/Photo Researchers, Inc., right Barry Singer/Animals Animals, bottom William R. Curtsinger/Photo Researchers, Inc.; 94 Ron Church/Photo Researchers, Inc., 95 Jeff Foott; 96–97 Norman Tomlin/Bruce Coleman, Ltd.; 98 top Fred Baldwin/Photo Researchers, Inc., bottom Kojo Tanaka/Animals Animals; 100 top left Michael and Barbara Reed/Animals Animals, right James D. Brandt/Animals Animals, center left and right Thase Daniel, bottom left and right Fred Bruemmer; 101 top George Holton, center Steven C. Wilson/Entheos, bottom François Gohier; 102 Jeff Foott; 104–105 Fred Bruemmer; 107 James H. Carmichael/Bruce Coleman, Ltd.; 108–109 Jeff Rotman; 111 David Doubilet/Animals, Animals; 112–113, 114 Douglas Faulkner; 115 top Bill Wood/ Natural History Photographic Agency, bottom Edward R. Degginger; 116 top left Howard Hall/Tom Stack and Associates, right Jacques Six; center left Charles Arneson; center right and bottom right Bill Wood/Natural History Photographic Agency; bottom left Jeff Rotman; 117 top Edward R. Degginger, center Jacques Six, bottom James H. Carmichael; 118 top Douglas Allan/Ecology Pictures, bottom James H. Carmichael; 119 Jeff Rotman; 120 top Douglas Faulkner, bottom Zig Lesczcynski/Animals Animals; 121 top Ron Taylor/Bruce Coleman, Inc., bottom Bill Wood/ Bruce Coleman, Inc.; 122–123 Ron Church/Photo Researchers, Inc.; 124 top Charles Arneson; 124 bottom, 125 Douglas Faulkner; 126 Jeff Rotman; 127 Phil Degginger; 128–129 Nicholas Devore/Bruce Coleman, Ltd.; 131, 132–133 Oxford Scientific Films; 134 top Robert Hermes/Photo Researchers, Inc., bottom Oxford Scientific Films; 135 Bruce Coleman, Ltd.; 136 Jane Burton/Bruce Coleman, Ltd.; 137 Jack Fields/Photo Researchers, Inc.; 138 Keith Gillett/Animals Animals; 140–141, 143 Bruce Coleman, Ltd.; 145 top Jane Burton/Bruce Coleman, Ltd., bottom Oxford Scientific Films; 146 Fred Baldwin/Photo Researchers, Inc.;

149 Steven C. Wilson/Entheos; 150 top Robert W. Hernandez/ Photo Researchers, bottom Bruce Coleman, Ltd.; 152–153 François Gohier; 154 top George Laycock/ Bruce Coleman, Ltd., bottom Russ Kinne/Photo Researchers, Inc.; 155 Thase Daniel; 156–157 Edward R. Degginger; 159 top Jen and Des Bartlett/Bruce Coleman, Ltd., center François Gohier, bottom Kojo Tanaka/Animals Animals; 160–161 Vancouver Aquarium—Row 1: left Stefani Hewlett, center and right Jeremy Fitz Gibbon; Row 2: left Jeremy Fitz Gibbon, center Hans DeJager, right N. A. Newman; bottom Stefani Hewlett; 162 top George Holton, bottom Joseph R. Jehl; 164–165 Philippa Scott/Photo Researchers, Inc.; 167 Peter David/Photo Researchers, Inc.; 168–169 Kenneth E. Lucas/Steinhart Aquarium; 171 top Peter David/Photo Researchers, Inc.; 171 bottom, 172–173 Oxford Scientific Films; 174 top Kenneth E. Lucas/Steinhart Aquarium; 174 bottom, 177 Peter David/ Photo Researchers, Inc.; 178–179, 180 Oxford Scientific Films; 183 Peter David/Photo Researchers, Inc.; 184, 186–187 Oxford Scientific Films; 189 Official Photography U.S. Navy; 190–191 Stephen J. Kraseman

APPENDICES

Life in the Seas

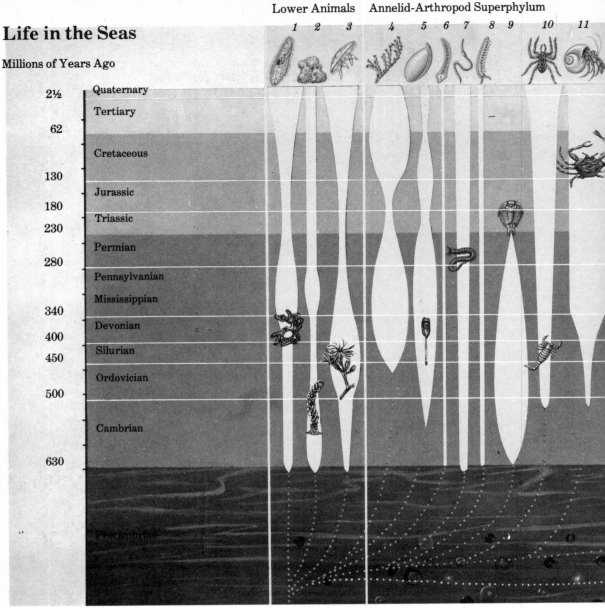

Millions of Years Ago

Lower Animals   Annelid-Arthropod Superphylum

1   2   3   4   5   6   7   8   9   10   11

Period	Millions of Years Ago
Quaternary	2½
Tertiary	62
Cretaceous	130
Jurassic	180
Triassic	230
Permian	280
Pennsylvanian	
Mississippian	
Devonian	340
Silurian	400
Ordovician	450
	500
Cambrian	
	630
Precambrian	

1. Protozoans
2. Sponges
3. Jellyfish, etc.
4. Moss Animals
5. Lampshells
6. Flatworms
7. Segmented Worms
8. Peripatus Worms
9. Trilobites
10. Spiders, Mites, etc.
11. Lobsters, Crabs, etc.
12. Insects
13. Mollusks
14. Starfish, etc.
15. Prevertebrates
16. Jawless Fishes
17. Placoderms
18. Cartilaginous Fishes
19. Bony Fishes
20. Amphibians
21. Reptiles
22. Birds
23. Mammals

194

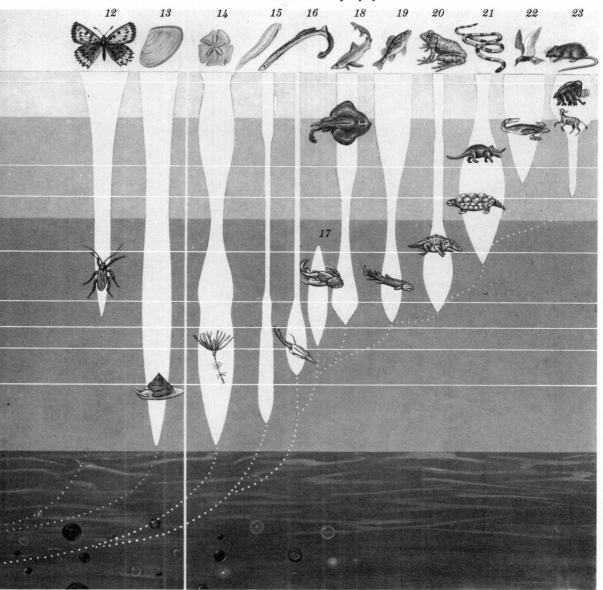

The millions of species of living organisms are divided into several major groups called phyla, 11 of which are shown here. Left to right: *single-celled animals (Protozoa); sponges (Porifera); polyps and jellyfish (Coelenterata); moss animals (Bryozoa); lampshells (Brachiopoda); flatworms (Platyhelminthes); segmented worms (Annelida); the insect-spider-crab complex (Arthropoda); mollusks (Mollusca); the sea star–crinoid complex (Echinodermata); the backboned animals (Chordata). Most fossilized remains, found in rocks of various ages, are members of phyla represented by species living today. Stratigraphic distribution—that is, according to layers of rock in the earth—is measured in units* called periods, of which there are 13. The oldest period, and the one of longest duration (over 4,000 million years), is the Precambrian. The other periods are of variable length, from 2 million years (Quaternary) to over 100 million years (Cambrian). All the periods together form the stratigraphic column, in which the oldest layers are at the bottom and the youngest are at the top. The names of the periods were first applied to systems of rocks, as exposed in different areas of the Earth, long before scientists could accurately estimate their ages. Prior to the nineteenth century, geologists estimated the entire stratigraphic column was formed during the last few thousand years. Modern geologists, after studies of rock radioactivity and other indica-tors, estimate the time to be nearly 5 billion years. Animals represented by abundant fossils generally have a skeleton; usually, it is only the skeleton, not the soft tissue, that fossilizes. The study of animal inter-relationships, as exemplified in the tree of animal life shown here, is known as zoological systematics.

Diatoms, the most abundant plants in the ocean, occur in the sunlit portion of the ocean, from the surface to a depth of 200 meters. There are about 600 different species of diatoms, in an amazing variety of forms. Some are planktonic and drift in the water; others are benthonic and live on the bottom. All diatoms are alike in having a silica ("glass") shell, with at least two parts, top and bottom. Through photosynthesis, diatoms transform energy from the sun into food. (The pigment, chlorophyll a, which manufactures food, is contained in the chromatophore.) Diatoms reproduce by cell division (mitosis). When a diatom divides into two daughter diatoms, one inherits the top shell of the parent and makes a new bottom shell. This daughter is the same size as the parent. The other daughter diatom, which inherits the bottom shell of the parent, is smaller than the parent and its sister diatom. With further cell division, the daughter diatoms become smaller and smaller, until they reach a critically small size. Then each escapes from its tiny shell and makes a new one of the normal larger size. The reduction process then begins again.

Phytoplankton/Diatoms

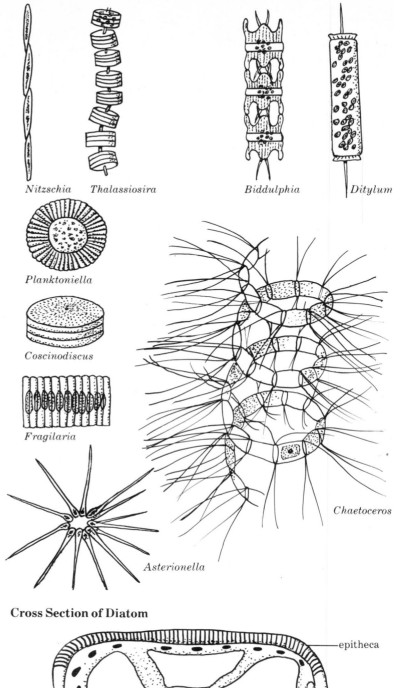

Nitzschia   Thalassiosira

Biddulphia   Ditylum

Planktoniella

Coscinodiscus

Fragilaria

Asterionella

Chaetoceros

Cross Section of Diatom

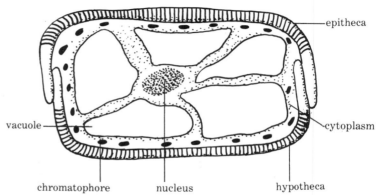

epitheca

vacuole

cytoplasm

chromatophore   nucleus   hypotheca

The plankton include both plants (phytoplankton) and animals (zooplankton). Those which are permanent drifters are called holoplankton. Other zooplankton which are drifters when young but which become bottom-dwellers or swimmers as adults are called meroplankton. Most meroplankton are larvae of bottom-dwelling invertebrate animals such as worms, echinoderms, crustaceans, and mollusks. A larva is a small, often microscopic life form that bears little resemblance to the adult stage and is adapted to a different way of life. Larvae feed on phytoplankton and smaller zooplankton until they metamorphose into juveniles, which resemble adults more closely. In many species, the larvae permanently attach to surfaces suitable for adult life. After settling, the larvae transform.

Zooplankton

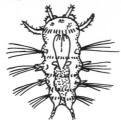

larvae—*Platynereis agassizi*—meroplankton

larvae—zoea of sand crab—*Emerita analoga*—meroplankton

pluteus larvae of sea urchin—meroplankton

veliger larva of snail—meroplankton

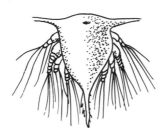

nauplius larva of barnacle—meroplankton

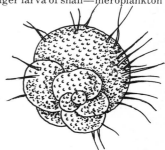

foraminifera—holoplankton

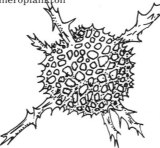

radiolarian—holoplankton

dinoflagellate—holoplankton

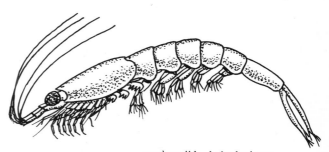

euphausiid—holoplankton

Algae

Multicellular marine algae, commonly called seaweeds, are plants of the intertidal and shallow zones of the ocean. Most of them, such as the rockweed Fucus, attach to rocky surfaces; but a few such as Sargassum drift with the plankton. The coralline algae, such as Lithothamnion, are hard and resemble rocks. They are important in the formation of coral reefs, for they flourish in the heavy surf on the windward edge of the reef, where few other organisms live. Algae are grouped according to color: for example, brown algae (Phaeophyta) and red algae (Rhodophyta). All algae have chlorophyll, and without the other pigments they would all be green. Algae never have flowers and seeds, but they nevertheless reproduce sexually by the union of gametes—cells comparable to eggs and sperms.

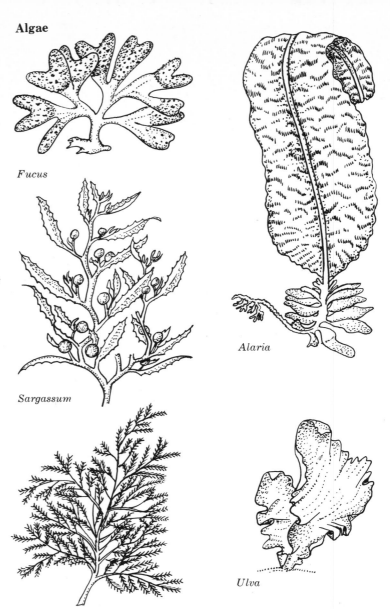

Fucus

Sargassum

Alaria

Polysiphonia

Ulva

Lithothamnion

Sponges are animals of the phylum Porifera ("pore bearers"). All are attached bottom-dwellers as adults, and until the eighteenth century they were thought to be plants. Then scientists discovered they are filter-feeders that pump water through numerous surface pores into a central cavity. As the water passes through the body wall of the sponge, food particles are removed and consumed. A sponge often resembles a small chimney, with a single large hole at the top. There are about 150 freshwater species; all the others, about 5,000 species, are marine. All have a skeleton that is a network of needle-like and sometimes branching elements called spicules, or of interlaced fibers. Sponges are divided into three groups: calcareous sponges (Calcispongiae), glass sponges (Hyalospongiae), and common sponges (Demospongiae). Bath sponges are the cleaned and dried skeletons of common sponges (family Spongiidae). Sponges reproduce sexually, and the fertilized egg develops into a swimming larva. The larva soon settles and transforms into a tiny sponge.

Sponges

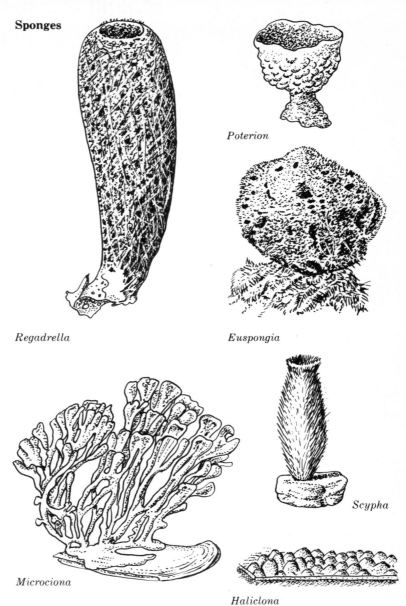

Regadrella

Poterion

Euspongia

Microciona

Scypha

Haliclona

Development of a Calcareous Sponge

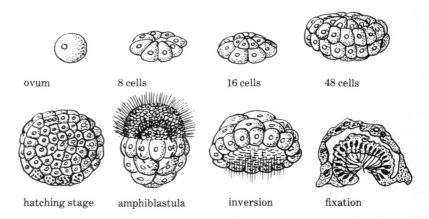

ovum

8 cells

16 cells

48 cells

hatching stage

amphiblastula

free-swimming

inversion

fixation

(seen in section)

Corals

Fungia

Acropora

Oculina

Meandrina

Tubipora

Coelenterates

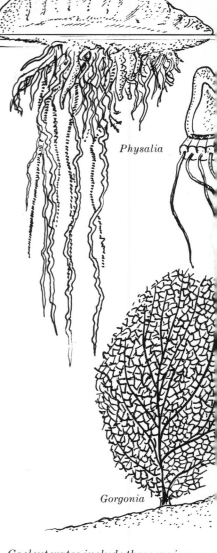

Physalia

Gorgonia

Cross Section Of Coral Polyp

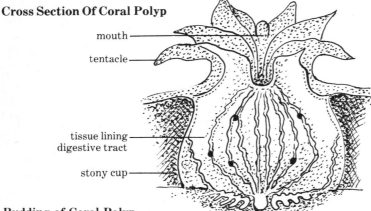

- mouth
- tentacle
- tissue lining digestive tract
- stony cup

Budding of Coral Polyp

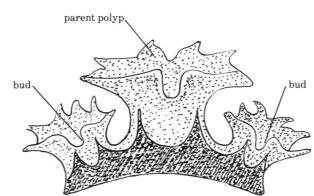

- parent polyp
- bud
- bud

Coelenterates include three major groups: hydroids (Hydrozoa), about 3,000 species; jellyfishes (Scyphozoa), about 200 species; anemones and corals (Anthozoa), about 6,000 species. Two basic body types occur—the polyp and the medusa. The polyp is typically an attached form, resembling a hydra, anemone, or coral. The medusa is typically a swimming form, resembling a jellyfish. Some species have both polyp and medusa stages in their life cycle. The 9,000 species of hydroids, jellyfishes, sea anemones, and corals make up the phylum Coelenterata (Cnidaria). Corals are either solitary or reef-building. Coral reefs, composed of many different coral colonies in association with other animals and plants, form massive structures the size of hills. The two

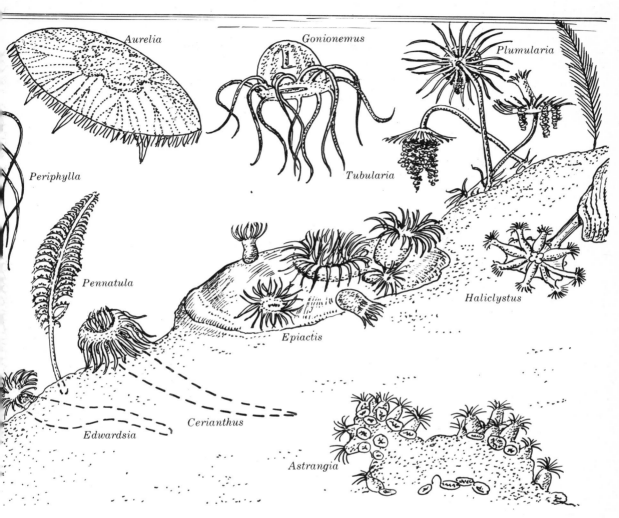

Aurelia
Gonionemus
Plumularia
Periphylla
Tubularia
Pennatula
Epiactis
Haliclystus
Edwardsia
Cerianthus
Astrangia

most prominent kinds of corals are the octocorals (Alcyonaria) and the stony corals (Scleractinia). Octocorals, which have eight tentacles around the mouth, include organ-pipe corals, soft corals, blue corals, gorgonians, red corals, and sea pens. All form colonies. Stony corals have more than eight tentacles; most form colonies. The living part of the coral is only a thin film of soft tissue over the surface of the hard skeleton. Related to stony corals are black corals (Antipatharia), which grow in deepwater in tropical regions and are sought by the jewelry trade.

Corals enlarge their colony by budding new polyps, each with its own mouth, tentacles, and digestive cavity; but all the polyps in a colony remain interconnected and share nutrition. Corals also reproduce sexually, by producing eggs and sperm. The fertilized egg develops into a larva, which soon quits the plankton, attaches, grows tentacles, and takes up the life of a polyp. By budding new polyps, the larva may eventually grow into a large coral colony of its own.

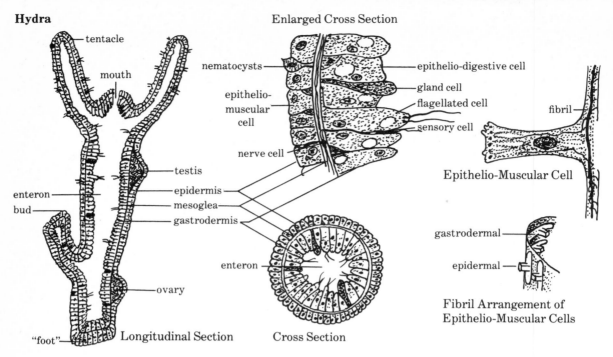

Hydra

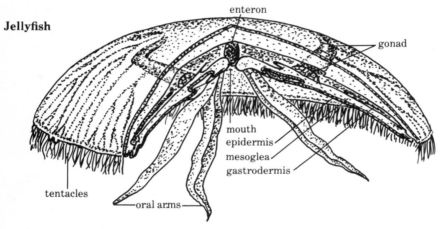

Jellyfish

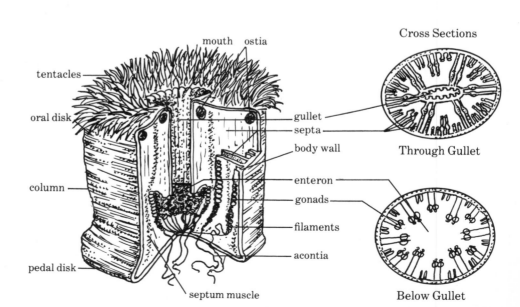

Sea Anemone

Coelenterates have a body wall with two layers of cells: an outer layer (epidermis) and an inner layer (gastrodermis). Between is a generally noncellular layer that is sometimes thin (hydra) and sometimes thick and jelly-like (jellyfish). Inside is a space (enteron) where food is digested. Various cells in the body wall perform different functions. Epithelio-muscular cells, for example, can contract, thereby changing the shape of the animal. A hydra extends itself by taking in water through the mouth, and in this relaxed state the animal may reach a length of 20 millimeters. Contraction of muscular cells shrinks the animal into a tight ball a half millimeter in diameter. Hydras relax and contract every 5 to 10 minutes. The jellyfish is more complex: it is shaped like an umbrella, with tentacles hanging from the margin, and the mouth in the undersurface. Contraction of muscular cells around the umbrella margin causes the umbrella to contract and propels the animal, usually upward, through the water. The sea anemone, like the coral, has a gullet that extends partway into the enteron. Anemones, like other coelenterates, have reproductive organs that release either eggs or sperm. Anemones sometimes reproduce by simply dividing in two, by budding off pieces from the top, or by leaving behind fragments of the pedal disk as the animals move away.

The marine hydroid Obelia includes both polyp and medusa stages in its life cycle. The polyp stage is colonial, and the colony resembles a tiny bush about 10 centimeters in height. Like coral polyps, hydroid polyps are interconnected and share a common nutrition. The colony grows by budding new polyps and stems. Obelia has an external skeleton or shell. There are always at least two kinds of polyps in colonial hydroids: polyps that feed and those which reproduce by budding medusae. The medusae, which resemble small jellyfishes, reproduce by producing eggs and sperm. The fertilized egg develops into a larva, which after settling transforms into a polyp. Through budding, a polyp can grow into a large colony.

Medusae — gonads — tentacles — sperm — ovum

Hydranths

expanded   contracted

mouth

gonangium

medusa buds

bud

coenosarc

horny perisarc

part of a mature colony

Sexual Reproduction

zygote

blastula

swimming

planula

Asexual Budding

starts new colony

settles

entire colony, life size

203

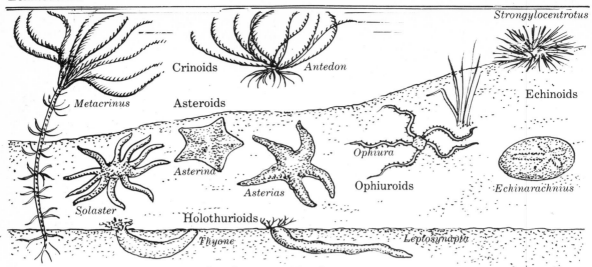

Metacrinus · **Crinoids** · *Antedon* · *Strongylocentrotus* · **Echinoids** · **Asteroids** · *Asterina* · *Ophiura* · **Ophiuroids** · *Echinarachnius* · *Asterias* · *Solaster* · **Holothurioids** · *Thyone* · *Leptosynapta*

The phylum Echinodermata ("spiny skins") includes about 6,000 marine species. The five major groups of living echinoderms are: sea stars (Asteroidea), brittle stars (Ophiuroidea), sea urchins (Echinoidea), sea cucumbers (Holothuroidea), and sea lilies and feather stars (Crinoidea). Most are bottom-dwellers, but feather stars are swimmers. As adults, echinoderms are radially symmetrical, resembling a circle subdivided into five parts. Despite their common five-rayed symmetry, echinoderms are entirely different from one another in gross appearance. Most species have a prominent skeleton in the form of spines and plates.

A sea star develops through the usual stages from egg to the first feeding stage—the larva. Some weeks later, projections (arms) begin to develop, and these enable the larva to attach to a surface and transform into an adult.

Development of Sea Star

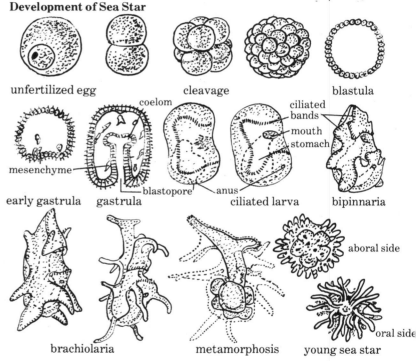

unfertilized egg · cleavage · blastula

coelom · ciliated bands · mouth · stomach · mesenchyme · blastopore · anus · early gastrula · gastrula · ciliated larva · bipinnaria

brachiolaria · metamorphosis · aboral side · oral side · young sea star

Sea Star

disk · madreporite · anus · spines · tentacle

arms · ambulacral grooves · tube feet · mouth

Aboral Surface · Oral Surface

Sea Urchin

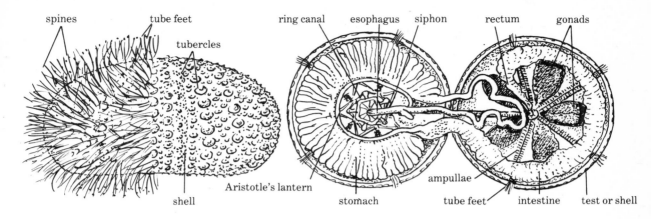

spines tube feet tubercles ring canal esophagus siphon rectum gonads

Aristotle's lantern shell stomach ampullae tube feet intestine test or shell

Sea Cucumber

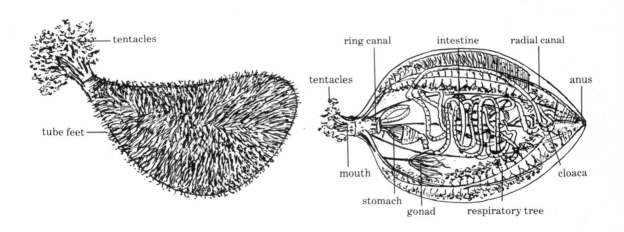

tentacles

tube feet

ring canal intestine radial canal

tentacles anus

mouth cloaca

stomach gonad respiratory tree

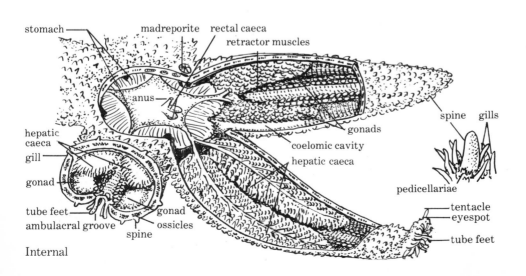

stomach madreporite rectal caeca retractor muscles

anus

spine gills

hepatic caeca gonads

gill coelomic cavity

gonad hepatic caeca

pedicellariae

tube feet gonad

ambulacral groove ossicles

spine

tentacle

eyespot

tube feet

Internal

Mollusks

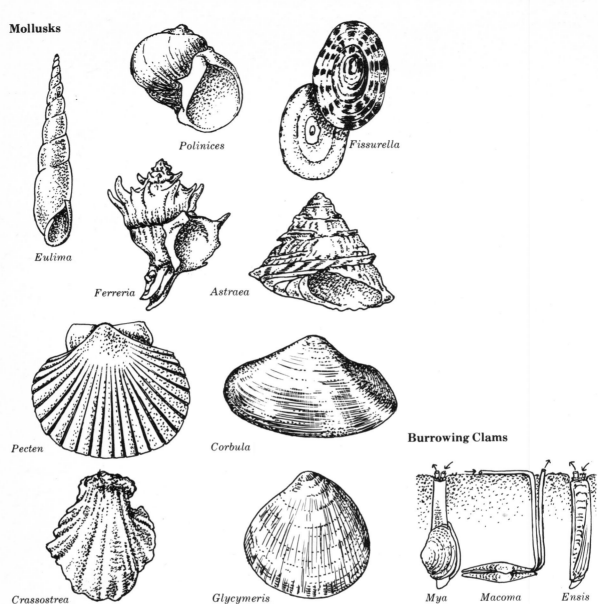

Eulima

Polinices

Fissurella

Ferreria

Astraea

Pecten

Corbula

Burrowing Clams

Crassostrea

Glycymeris

Mya   Macoma   Ensis

Members of the phylum Mollusca are among the most numerous and conspicuous members of the marine community. There are about 80,000 living species of mollusks, most of which are marine, and an additional 40,000 species are known through fossils. Typically, a shell is present. Mollusks are structurally complex, with separate digestive, nervous, circulatory, reproductive, and locomotor systems. The basic structure is best represented by chitons. The shell of the chiton, with its 8 overlapping plates, is a more useful, flexible protective cover than was the solid dome of a shell borne by its ancient ancestor. The major groups of living mollusks include: snails (Gastropoda), with 40,000 species; primitive mollusks (Monoplacophora),

with only 7 species, the first discovered in 1952; chitons (Polyplacophora), with 600 species; solenogasters (Aplacophora), with 130 species; clams and oysters (Bivalvia), with 40,000 species; tooth shells (Scaphopoda), with 200 species; squids and octopuses (Cephalopoda), with 200 species. Of the Cephalopods, only nautilus has an external shell. Despite the seeming burden of the large, beautifully patterned shell, nautilus can swim with surprising rapidity.
Bivalves are called filter-feeders. Many burrow into the bottom and, by means of inflowing and outflowing tubes (siphons), filter food particles from the steady stream of water that moves through them.

206

Cephalopods

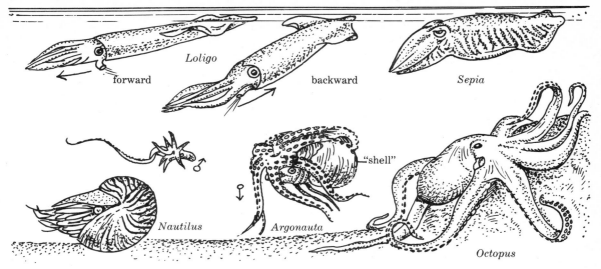

forward   *Loligo*   backward   *Sepia*

Nautilus   *Argonauta*   "shell"   *Octopus*

Chiton

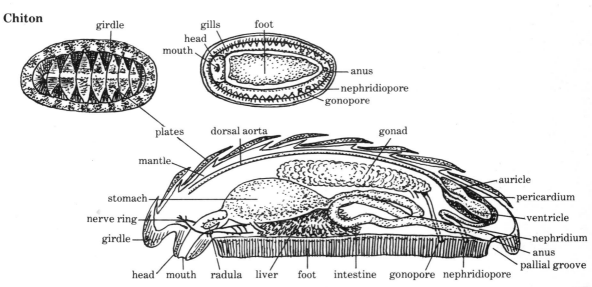

girdle   gills   foot   head   mouth   anus   nephridiopore   gonopore

plates   mantle   dorsal aorta   gonad   auricle   pericardium   ventricle   nephridium   anus   pallial groove

stomach   nerve ring   girdle   head   mouth   radula   liver   foot   intestine   gonopore   nephridiopore

Nautilus

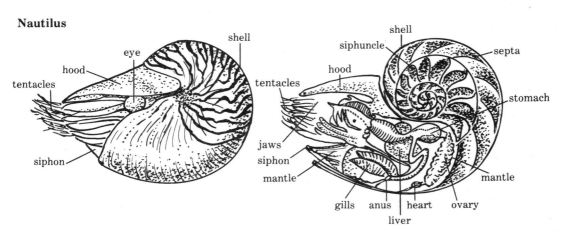

eye   shell   hood   tentacles   siphon

siphuncle   shell   septa   hood   tentacles   stomach   jaws   siphon   mantle   gills   anus   heart   ovary   liver   mantle

Crustaceans

Crustaceans are members of the phylum Arthropoda ("jointed feet"), of which most species, numbering in the millions, are terrestrial. Terrestrial arthropods include the insects, spiders, and mites. With a mere 30,000 mostly marine species, crustaceans might seem a relatively small group, but there are probably more crustaceans in the ocean than any other multicellular animals. Like other arthropods, crustaceans are segmented, and the segments bear paired and jointed appendages. In crustaceans, the segments are modified into three main body regions: head, thorax, and abdomen. And the appendages are modified into antennae, mandibles, maxillae, and legs. There are six major groups of crustaceans: primitive crustaceans (Cephalocarida), with only 4 species, the first discovered in 1955; fairy shrimps (Branchiopoda), with 800 species; mussel shrimps (Ostracoda), with 2,000 species; copepods (Copepoda), with 4,500 species; barnacles (Cirripedia), with 1,000 species; and all the others (Malacostraca), with 20,000 species. The many subgroups of malacostracans include the familiar shrimps, lobsters, and crabs (Decapoda), with 9,000 species. Other important malacostracans are isopods (Isopoda), with 4,000 species, and amphipods (Amphipoda), with 5,000 species.

The life of a crustacean begins with a fertilized egg, which typically develops into a series of planktonic larval stages. In the shrimp Penaeus, *there are four different stages. Crustaceans, like other arthropods, have an external and chitinous skeleton that limits the growth of the body within. Periodically, the old skeleton is molted. Larvae also molt as they grow.*

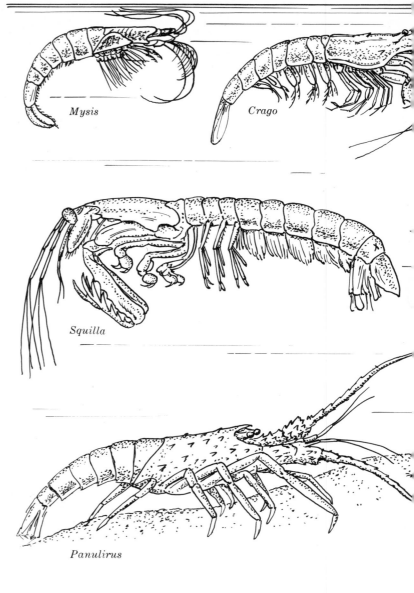

Mysis

Crago

Squilla

Panulirus

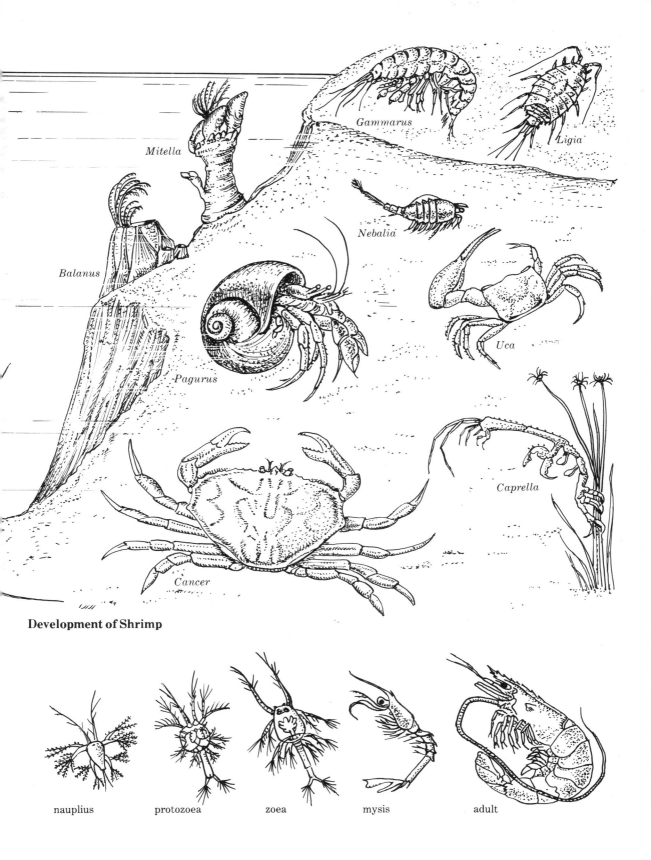

Gammarus

Ligia

Mitella

Nebalia

Balanus

Uca

Pagurus

Caprella

Cancer

Development of Shrimp

nauplius   protozoea   zoea   mysis   adult

Evolution of Fishes

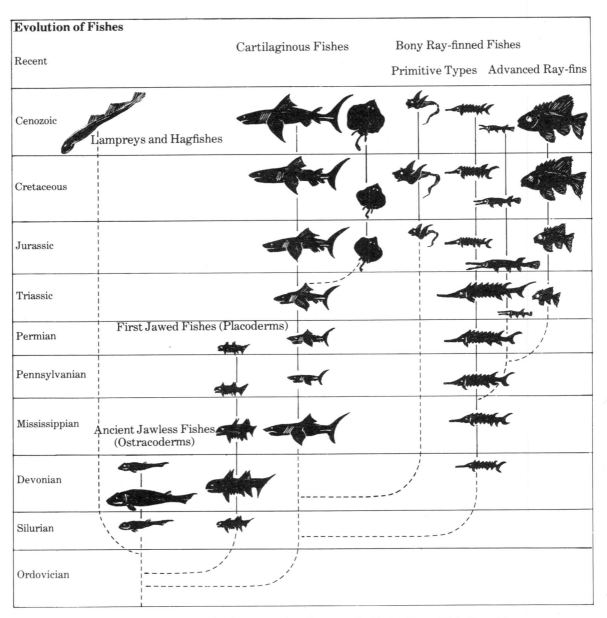

Recent

Cartilaginous Fishes

Bony Ray-finned Fishes

Primitive Types   Advanced Ray-fins

Cenozoic

Lampreys and Hagfishes

Cretaceous

Jurassic

Triassic

First Jawed Fishes (Placoderms)

Permian

Pennsylvanian

Mississippian

Ancient Jawless Fishes
(Ostracoderms)

Devonian

Silurian

Ordovician

Fishes comprise about one-half of the 50,000 species of the phylum Chordata. The others are the mainly land-dwelling amphibians, reptiles, birds, and mammals. There are three major groups of living fishes: lampreys and hagfishes (Cyclostomata), with a few dozen species; the sharks, rays, and chimaeras (Chondrichthyes), with 1,000 species; and the ray-finned fishes (Actinopterygii), with 25,000 species. All but a few actinopterygians belong to the group Teleostei and are called "teleosts" for short.

Bony Fish (Perch)

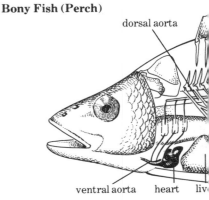

dorsal aorta

ventral aorta   heart   liv

stor

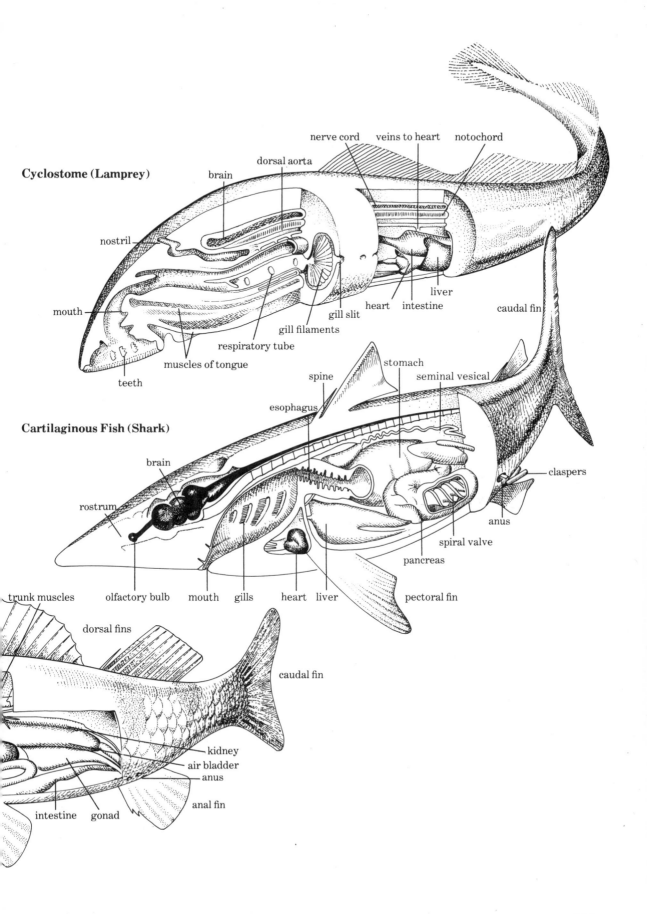

Cyclostome (Lamprey)

nerve cord

veins to heart

notochord

dorsal aorta

brain

nostril

mouth

gill slit

heart

intestine

liver

caudal fin

gill filaments

respiratory tube

muscles of tongue

teeth

Cartilaginous Fish (Shark)

spine

stomach

seminal vesical

esophagus

brain

claspers

rostrum

anus

spiral valve

pancreas

olfactory bulb

mouth

gills

heart

liver

pectoral fin

trunk muscles

dorsal fins

caudal fin

kidney

air bladder

anus

intestine

gonad

anal fin

A shark has an internal cartilaginous skeleton, the elements of which are nevertheless calcified and sometimes become as hard as bones. The brain is extremely primitive (the cerebrum is greatly reduced); thus the shark acts "blindly." Contained in the fore part of the intestine is a spiral valve which slows the passage of food, thereby increasing the amount of food absorbed. Its pectoral fin helps stabilize the fish as it swims. Sharks are ever-present, even in areas where humans bathe without incident. Most species do not normally molest human bathers and swimmers but are potentially dangerous if provoked. A few species are dangerous even if unprovoked. The most fearsome is the great white shark (Carcharodon), which has been implicated, through identifiable tooth fragments left in its victims' wounds and in the wood of small boats, in many unprovoked attacks upon humans.

Most sharks bring forth their young alive, but some lay eggs. The egg cases, called mermaid purses, sometimes wash up on the beach, and may even have a living embryo inside them.

Sharks

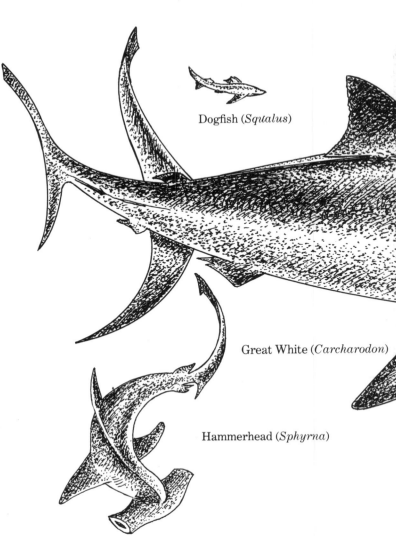

Dogfish (*Squalus*)

Great White (*Carcharodon*)

Hammerhead (*Sphyrna*)

Egg Cases

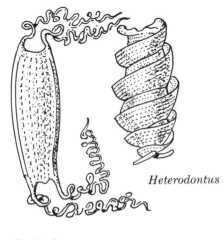

Scyliorhinus

Heterodontus

Pectoral Fin in Sections

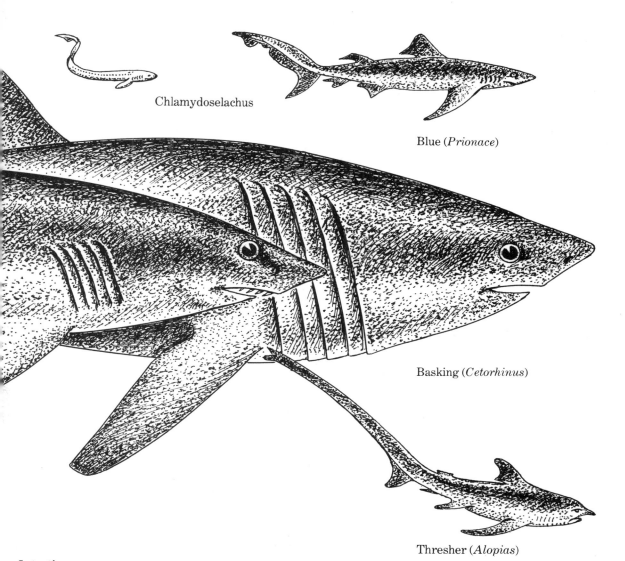

Chlamydoselachus

Blue (*Prionace*)

Basking (*Cetorhinus*)

Thresher (*Alopias*)

Intestine

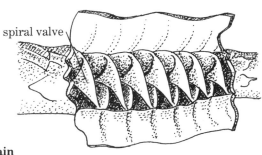

spiral valve

Brain

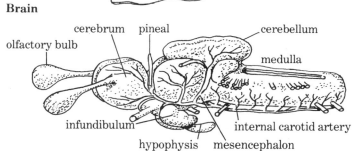

olfactory bulb

cerebrum   pineal   cerebellum

medulla

infundibulum

internal carotid artery

hypophysis   mesencephalon

Teleosts

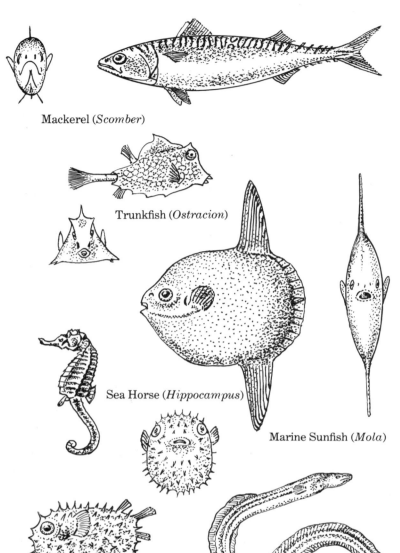

Mackerel (*Scomber*)

Trunkfish (*Ostracion*)

Sea Horse (*Hippocampus*)

Marine Sunfish (*Mola*)

Globefish (*Chilomycterus*)

Common Eel (*Anguilla*)

Teleostean fishes number about 25,000 species and include several major groups: bony-tongues and elephant fishes (Osteoglosso morpha), with 150 species; ladyfishes and eels (Elopomorpha), with 700 species; herrings and anchovies (Clupeomorpha), with 300 species; characins, minnows, and catfishes (Ostariophysi), with 6,000 species; salmons, trouts, and pikes (Salmoniformes), with 250 species; cods, toadfishes, and anglerfishes (Paracanthopterygii), with 1,000 species; and spiny-finned fishes (Acanthopterygii), with 15,000 species. The great variety of form and adaptation of teleosts belies the impression that all fishes are basically similar and primitive forms of life because they live in water and breathe with gills. The reality is otherwise. Some fishes periodically leave the water; many others breathe air and, under some circumstances, cannot survive without it. With respect even to brain structure, there are certain fishes (family Mormyridae) in which the brain is among the most complicated of those of all living organisms. Bony fishes breathe as water is pumped through the mouth and over the gills by the flapping action of the operculum or gill cover. In sharks, on the other hand, water is moved through the mouth and over gills as the shark swims.

There are 3 basic types of scales: ganoid scales, found on primitive fishes such as the sturgeons; cycloid scales, found on cod and haddock; and ctenoid scales, found on flatfishes such as flounders and soles.

Heterocercal tails of primitive fishes, such as sturgeon have dissimilar lobes in terms of size and shape (the lower lobe is shorter than the top). However, homocercal tails of more advanced fishes, such as cod or salmon have lobes that are equal in size and shape.

Fishes are colored by pigments in cells called chromatophores ("color bearers"). If the pigment is concentrated in a small area within the cell, it is not visible except as a dot. If the pigment is dispersed throughout the cell, it is visible as general color. In many fishes the cells are under nervous control, so that the pigments may be rapidly dispersed or concentrated in quick changes of color.

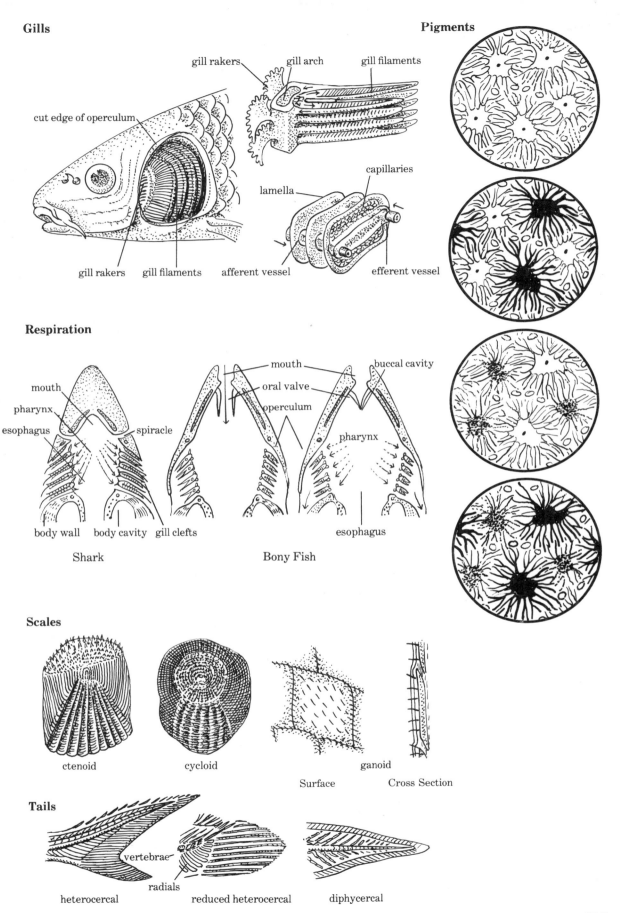

Gills

gill rakers
gill arch
gill filaments
cut edge of operculum
capillaries
lamella
afferent vessel
efferent vessel
gill rakers
gill filaments

Pigments

Respiration

mouth
oral valve
operculum
buccal cavity
mouth
pharynx
esophagus
spiracle
pharynx
body wall
body cavity
gill clefts
esophagus

Shark

Bony Fish

Scales

ctenoid
cycloid
ganoid
Surface
Cross Section

Tails

vertebrae
radials
heterocercal
reduced heterocercal
diphycercal

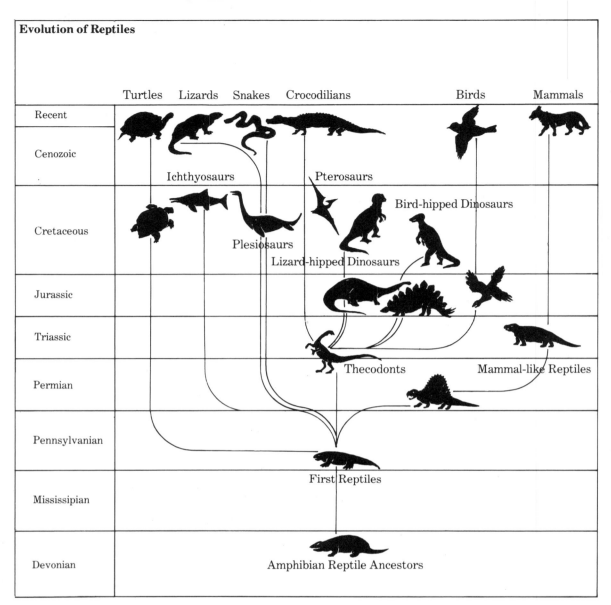

Evolution of Reptiles

	Turtles	Lizards	Snakes	Crocodilians		Birds	Mammals
Recent							
Cenozoic							
Cretaceous							
Jurassic							
Triassic							
Permian							
Pennsylvanian							
Mississipian							
Devonian							

Ichthyosaurs

Pterosaurs

Bird-hipped Dinosaurs

Plesiosaurs

Lizard-hipped Dinosaurs

Thecodonts

Mammal-like Reptiles

First Reptiles

Amphibian Reptile Ancestors

Land-dwelling chordates (Tetrapoda) include two basic groups: frogs and salamanders (Amphibia), with 2,500 species; reptiles, birds, and mammals (Amniota), with 18,000 species. Whereas amphibians typically lay their eggs in water, amniotes either lay a shelled egg on land or retain the egg in the female's body and give birth to living young. There are no marine amphibians, and salt water is fatal to amphibian eggs. Amniote eggs are protected, either ashore or inside the female, and all the amniote groups have marine representatives. There are marine species of turtles, lizards, snakes, crocodiles, birds, and mammals. Birds (Aves) and mammals (Mammalia) are the only warm-blooded chordates, but the birds are more closely related to crocodiles, which are cold-blooded, than to any other living animals.

Plesiosaurs, extinct Jurassic reptiles, reached a length of 15 meters. Their arms and legs were modified as flippers.

Early land-dwellers were fishes that walked from the water on primitive arms and legs. The oldest-known land-dweller is Ichthyostega, a Devonian fossil found in Greenland. Ichthyostega had a fish-like tail, complete with fin-rays.

Mosasaurs, extinct Cretaceous reptiles, were fish-eaters that lived in the ocean. Some grew to 10 meters in length.

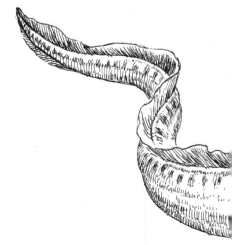

Plesiosaurus

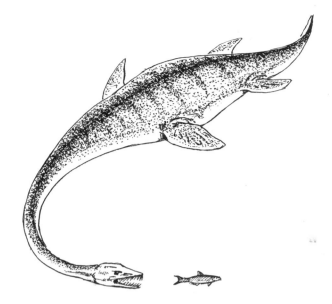

Ichthyostega

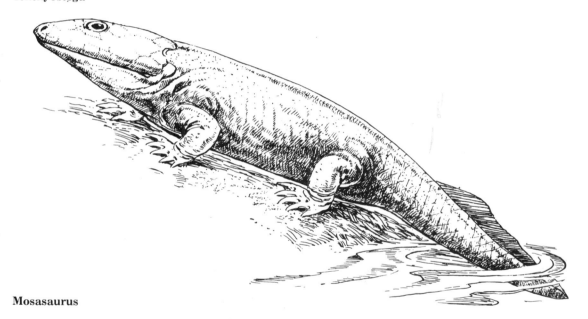

Mosasaurus

217

Leatherback (*Dermochelys*)

Loggerhead (*Caretta*)

Green (*Chelonia*)

The dozen or so species of marine turtles (families Dermochelidae and Cheloniidae) are all in danger of extinction because of human interference with their reproductive activities. Many governments now protect sea turtles by laws regulating or prohibiting their capture; but such laws are not always obeyed, because they cannot be efficiently enforced. In the remote areas where they come ashore, egg-laying females are too easily hunted with impunity by local inhabitants.

Head Shields

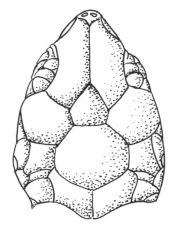

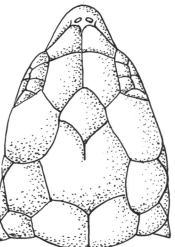

Chelonia

Eretmochelys

Hawksbill (*Eretmochelys*)

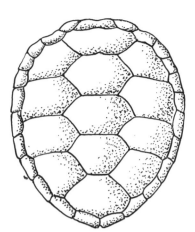

Chelonia/Eretmochelys

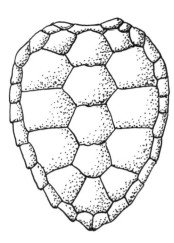

Caretta

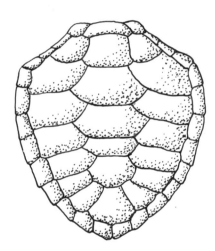

Lepidochelys

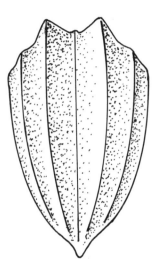

Dermochelys

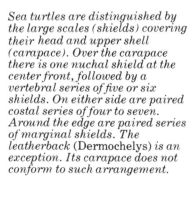

Sea turtles are distinguished by the large scales (shields) covering their head and upper shell (carapace). Over the carapace there is one nuchal shield at the center front, followed by a vertebral series of five or six shields. On either side are paired costal series of four to seven. Around the edge are paired series of marginal shields. The leatherback (Dermochelys) *is an exception. Its carapace does not conform to such arrangement.*

Caretta

Whales

Whales and dolphins comprise the mammalian group Cetacea, with about 90 species, the larger of which are in danger of extinction. Whaling, because of the impending extinction of whales, is now regulated through international agreement between the countries (Japan and the Soviet Union) that continue to engage in such commercial activity.

Sperm (*Physeter*)

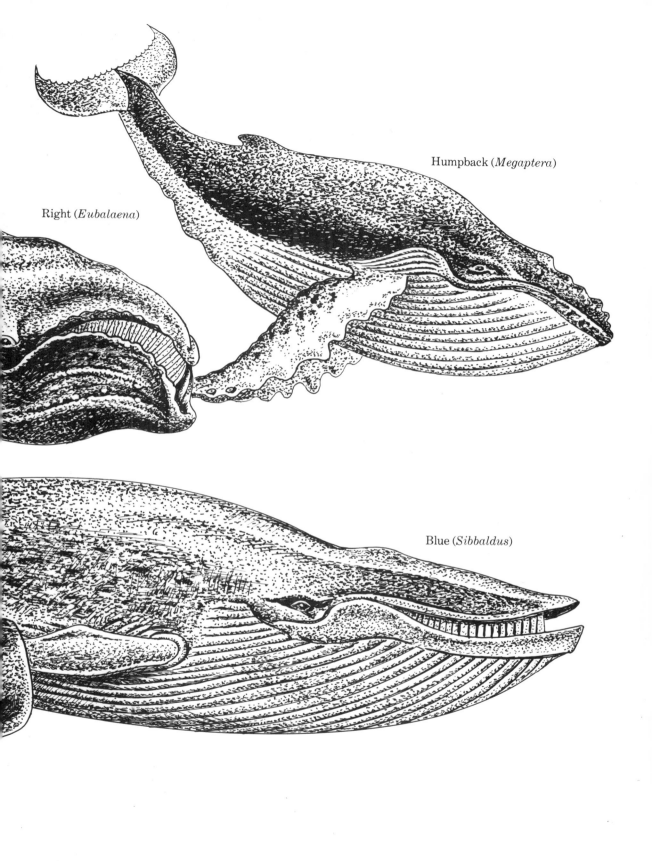

Humpback (*Megaptera*)

Right (*Eubalaena*)

Blue (*Sibbaldus*)

Glossary

Abyss. The deep sea, from about 2,500 to 6,000 meters, where the water temperature does not exceed 4°C. "Abyssal" is a term describing marine organisms especially adapted to the great pressure, low temperature, and lack of light at these depths.

Abyssal plain. Gently sloping or nearly level region of the sea floor in the deep, or abyssal, region. Long thought to be a featureless area devoid of life, it is now known to have ridges, hills, rifts, and a variety of marine organisms, including fishes.

Abyssopelagic. Pertaining to that part of the water (as distinct from the ocean floor) in the abyss.

Anadromous. Pertaining to fishes that live in the sea but spawn in fresh or brackish waters. Common examples are the salmons and the striped bass (*Morone saxatilis*). The salmons make spectacular migrations from the sea up freshwater rivers and streams to find spawning sites.

Atoll. A ring-shaped reef with a central lagoon. Beginning as a fringing reef around a volcano, the reef progresses through stages over geologic time. The atoll is one of the final stages.

Azooic zone. Literally, a zone without life. For centuries, men thought the ocean lacked life below a certain depth. But life in some form has been found at all depths, including in the deepest trenches.

Backshore. A part of the beach that is usually dry and becomes wet only at the highest tide (see *Spring tide*).

Barrier reef. A reef separated from a nearby landmass by a lagoon. The most famous is the Great Barrier Reef of Australia.

Basin. A depression in the sea floor. Usually it has little flow of water, resulting in near-stagnant conditions, with depleted oxygen and sometimes accumulations of organic matter and noxious gases such as hydrogen sulfide. Basins with adequate water flow and ample food have concentrations of marine organisms.

Bathyl. Pertaining to the ocean floor between 200 and 3,700 meters deep; approximately the same as the continental slope.

Bathypelagic. Pertaining to a zone in the sea (not the sea floor) between about 1,000 and 3,700 meters. It is a zone of perpetual cold (about 4°C) and perpetual darkness.

Bathythermograph. An oceanographic instrument that records water temperature and depth, but only in depths of less than 300 meters. Used to help trace water masses through temperature and to measure temperature of fish habitats.

Benthic. Pertaining to the marine environment of organisms that live permanently in or on the bottom. Such organisms are collectively termed the "benthos."

Biomass. The amount of living organisms in a given area, expressed as weight, or volume, of organisms per unit area or volume of the environment.

Brackish. Pertaining to water of less than normal marine salinity, that is, less than 32 to 35 °/oo. Brackish water is most common where rivers enter the sea and usually ranges between 0.5 and 17 °/oo. Such waters are favored by many species of shellfish, including clams and oysters, and by many fishes.

Catadromous. Pertaining to fishes that live in fresh water but spawn in the sea. The most famous examples are the eels (*Anguilla*), the American and European species of which migrate down rivers and streams to the sea, then swim hundreds of kilometers to a part of the Atlantic Ocean called the Sargasso Sea, where they spawn and then die.

Commensalism. A symbiotic relationship between two species in which one species is benefited and the other is not harmed. An example is the relationship between sharks and the remoras and shark suckers (Echeneidae), where the remora attaches itself to the shark by means of a sucking disk and is carried about. The remora does not harm the shark but is transported about and gets food from leftover scraps of the shark's feeding (see also *Mutualism*).

Continental drift. A term proposed by the geologist Alfred Wegener in 1912 to describe his theory that the continents, once part of a single landmass, have moved away from each other. In the eartly 1960s, theories of sea-floor spreading along mid-ocean ridges in the sea floor and the concept of plate tectonics furnished support for Wegener's theory.

Continental shelf. Zone adjacent to the continents or islands extending from the low-water mark to a depth arbitrarily set at 200 meters. Because the shelves are relatively shallow, receive considerable sunlight, and are generally swept by nutrient-bearing currents, they contain abundant populations of marine organisms and support most of the major world fisheries. Fishing grounds such as the Dogger Bank in the North Sea and the Grand Bank of Newfoundland and the Georges Bank in the northwest Atlantic Ocean are famous productive areas on continental shelves.

Coriolis effect. An effect on moving particles traveling over a rotating sphere (earth) whereby the particles are deflected to the right in the Northern Hemisphere and to the left in the Southern Hemisphere. The Coriolis effect has great influence on the movement of the major ocean currents, including the Gulf Stream/North Atlantic Current, thus modifying the climate and weather of coastal landmasses.

Currents. Horizontal movement of the water of the sea. Currents move water that is of favorable temperature and salinity for many marine organisms. They also transport eggs and larvae, as well as adult plankton. The most famous currents are the Gulf Stream, the Humboldt Current, and the Kuroshio.

Cyclone. A large storm rotating about an atmospheric low. The wind flow is counterclockwise in the Northern Hemisphere and clockwise (anticyclone) in the Southern Hemisphere. Cyclones include hurricanes and typhoons (tropical cyclones), as well as major storms (extratropical cyclones).

Dredge. A device for collecting benthic organisms. One type is dropped on a wire from a vessel to scoop up a sample of sea floor; another is towed over the bottom to collect materials from a large area. Dredges are also used in commercial shellfisheries for harvesting clams, oysters, and sea scallops.

Ecology. Science that studies the relationship between living organisms and their environment. Ecology includes the study of ecosystems such as a coral reef (in which scientists consider fishes, coral polyps, water currents, temperature and salinity and how all these factors interact).

Epifauna. Benthonic animals that live on the ocean floor, including scallops, oysters, and some species of sea urchins (see also *Infauna*).

Epipelagic. Pertaining to a part of the ocean from the surface to a depth of 200 meters, including the water in the shallows zone over the continental shelf.

Estuary. A tidal bay formed by the submergence of a river mouth or the lower part of a river valley. Well-known estuaries are found at the mouth of the Thames and Hudson rivers.

Euphotic zone. The uppermost layer of a body of water, which receives enough sunlight for photosynthesis to take place. Primary productivity occurs in this zone. Since there is no photosynthesis below the euphotic zone, organisms below that level must depend on consuming other organisms.

Euryhaline. Pertaining to certain organisms that can exist within a wide salinity range. Some killifishes (*Fundulus*) move freely from bay salinity (about 28 to 30 °/oo) to brackish salinities (less than 17 °/oo). Some fishes appear to be euryhaline when they are found well up a river; in actuality, however, they usually have remained in the saltwater wedge that advances upstream with the incoming tide (see also *Stenohaline*).

Fiord (or fjord). A narrow, deep, steep-sided inlet of the sea; usually a long, submerged glaciated valley along a mountainous coast. Although the most famous examples of fiords are in Norway, they also occur on the Pacific coast of Canada.

Fringing reef. A reef attached directly to a landmass, which is one of the first stages in the development of coral reefs leading to an atoll.

Fry. Young fish, hatchlings, or fingerlings; the stage after the larval stage when the fish have resorbed the yolk sac and are swimming and searching for food.

Groundfish. Fish that live on or near the sea floor. Examples include cod (*Gadus*), haddock (*Melanogrammus*), flounder (*Pseudopleuronectes*), sole (*Solea*), halibut (*Hippoglossus*), and pollock (saithe; *Pollachius*). Groundfish are among the most valuable fishes in commercial fisheries.

Guyot. A tablemount; a seamount with a comparatively smooth, flat top. Probably of volcanic origin, guyots were discovered and identified in the Pacific Ocean in World War II; though most common in that ocean, they occur throughout the world. The flat upper surface, which is as much as 1 kilometer below sea level, sometimes shows remains of reef-type corals (see also *Seamount*).

Hadal. Pertaining to the greatest depths of the oceans, in the deep-sea trenches, usually between 6,000 and 10,800 meters. Hauls with collecting dredges and, more recently, direct observations from manned submersibles have demonstrated the presence of fish and other living organisms in the hadal zone.

Hurricane. A severe, powerful tropical cyclone in the North Atlantic, Caribbean Sea, Gulf of Mexico, or eastern North Pacific. Winds in hurricanes exceed 120 kilometers per hour, accompanied by very heavy rainfall. The winds produce mountainous waves and are extremely destructive at sea and along coastal areas (see also *Cyclone; Typhoon*).

Infauna. Benthic animals that burrow or bore into the sea floor, or substrate. The best-known examples are the clams (*Mercenaria*) and bloodworms (*Glycera*).

Inquilinism. A symbiotic relationship in which one partner seeks refuge in the other; for example, the clownfish (*Amphiprion*) taking refuge in large sea anemones (*Stoichactis*) in the Indo-Pacific realm. There appears to be mutual benefit, with no harm to either partner. The clownfish attracts prey to the anemone and may also clean debris and wastes from its host. The anemone, in turn, provides refuge for the clownfish, and it may also provide scraps of food for the little fish.

Internal wave. A wave occurring within the sea, usually at the boundary between water masses of different density, as a result of different temperatures or salinity. The waves may be detected by submersibles passing through them or by measuring devices lowered from oceanographic vessels.

Island arc. A curved group of volcanic islands, commonly convex toward the open ocean. Often occurs on the landward side of a deep-sea trench. Best-known examples are the Aleutian Islands of Alaska and the islands of Japan. Such areas frequently are sites of volcanic and seismic activity, with resulting tsunamis (tidal waves).

Lagoon. A shallow body of water separated from the sea by a barrier island, reef, or similar feature.

Magma. Molten or fluid material within the earth from which igneous rock results by cooling. Magma escaping from rifts in the seabed adds to the sea floor and influences the spreading of the sea floor.

Mediterranean (when spelled lowercase). An oceanographic term referring to any sea that is, like the Mediterranean Sea, surrounded by land and has an opening to other bodies of water. Other mediterranean seas are the Caribbean and the Sea of Japan.

Mesopelagic. That part of the sea from 200 to 1,000 meters deep; sometimes called the "twilight zone." It is inhabited by animals which prefer the dark, and whose upward movements are limited by the penetration of sunlight and whose downward movements are limited by the cold water layers below (about 10°C and colder). These animals often migrate to the surface at night to feed, then return to the dark depths after dawn.

Monsoon (from Arabic, meaning "a season"). Name for seasonal winds. First applied to winds over the Arabian Sea that flow for six months from the northeast and for six months from the southwest, monsoon is now also used for similar winds in other parts of the world. Such winds from the sea bring much-needed rains.

Mutualism. A symbiotic relationship in which both partners benefit, with no harm to either. A common example is the sponges or sea anemones attached to the upper shell (carapace) of some crabs. The attached animals help to camouflage the crab and obtain some of their food from scraps lost by the crab (see also *Commensalism*).

Nansen bottle. A device used by oceanographers to collect water samples. It is lowered into the sea on a wire, with both ends open. At the desired depth, a metal messenger sent down on the wire trips the bottle, thereby closing both ends and trapping the water sample.

Neap tide. The lowest tide, occurring twice in each lunar month. It is the result of the sun and moon acting at right angles to the Earth (see also *Spring tide*).

Nekton. Animals in the pelagic environment that are active swimmers. The group includes fishes, reptiles (sea snakes, turtles), and marine mammals (whales, porpoises).

Oceanic. That portion of the pelagic environment seaward from the edge of the continental shelf or shallows zone; also called the "open sea."

Oceanography. Scientific study of the sea, including three main divisions: biological oceanography, the study of the plants and animals of the sea; physical oceanography, the study of the waves, currents, pressure, temperature, and other forces in the sea; and chemical oceanography, the study of the various chemical elements in the sea and their interrelationships.

Patch reefs. Small coral reefs without lagoons that may form part of a barrier reef or atoll rim.

Pelagic. Pertaining to the primary division of the sea, including the whole mass of water in the shallows and open ocean.

Photosynthesis. The process in green plants whereby sugars and starches are produced from carbon dioxide and water in the presence of sunlight. This is the primary process in the production of organic material (that is, food) from inorganic materials. It is an important function of algae, such as diatoms and kelps, and of marine grasses.

Plankton. Organisms, including both plants and animals, that are drifters or feeble swimmers and are generally small.

Plate tectonics. The theory that the earth's crust and upper mantle consist of large blocks or plates in constant motion. The movement is caused by magma rising out of the earth's molten core between adjacent plates. When plates collide, one is forced under the other; this usually occurs in the vicinity of oceanic trenches and island arcs.

Productivity. The rate at which photosynthesis takes place.

Reef. An organic, wave-resistant mass in the sea. In some places, any shallow hazard to navigation is considered a reef, regardless of origin.

Reversing thermometer. A mercury-in-glass thermometer that records temperature when inverted, and retains the reading until returned to its original position. These thermometers are extremely accurate (to 0.01°) and are used with Nansen bottles to determine the water temperature at predetermined depths.

Rip current. A usually strong and narrow current caused by the seaward flow of water piled up near shore by incoming waves. Such currents afford very unstable environments for marine animals because of the constant movement of the substrate.

Ripples (or ripple marks). Undulating features produced in sand by waves or currents. Oceanographers make observations of ripple marks on the sea floor, even at great depths, to estimate the direction and force of ocean currents. Fossilized ripple marks formed millions of years ago provide clues to the nature of ocean currents in seas long disappeared.

Running wave. The common wave propagated across a water surface. The height of the wave, the distance from crest to crest, and whether or not it breaks depend on the wind speed, the fetch (distance the wind travels over the open sea), and the water depth.

Salinity. The quantity of dissolved solids in seawater; specifically, the total amount of dissolved solids per unit volume of seawater, expressed as parts per thousand ($^o/oo$). Open ocean seawater has about 32 to 35 $^o/oo$. Salinity is useful in identifying water masses that may occur as "bubbles" of low-salinity water within a greater mass of high-salinity water. Salinity is also a factor limiting the abundance and distribution of many marine organisms.

Salinometer. Any device used for determining the salinity of water.

SCUBA. *S*elf-*C*ontained *U*nderwater *B*reathing *A*pparatus. Unlike the bulky, hard-hat diving dress that ties the diver to the mother ship with various hoses and lines, scuba allows the diver to move about freely at limited depths. The tanks of the apparatus, commonly containing compressed air, limit the diver to depths of less than 100 meters.

Sea-floor spreading. A general term covering the growth of the sea floor as magma rises along the mid-ocean ridges and spreads laterally away from them. Sea-floor spreading is the major mechanism on which the hypothesis of continental drift and plate tectonics is based.

Seamount. An undersea, narrow-topped mountain, usually rising 1,000 meters or more from the sea floor.

Sessile benthos. Bottom-dwelling organisms attached to the sea floor. They have very little or no mobility.

Shallows zone (see *Continental shelf*).

Spring tide. The highest tide in the lunar cycle, occurring twice in each lunar month—when the Earth, sun, and moon are aligned (see also *Neap tide*).

Stenohaline. Pertaining to certain organisms that can exist only within a narrow salinity range (see also *Euryhaline*).

Strand line. The shoreline; also the line of debris washed up onto the beach by waves. "Stranding" is a term used to describe the state of objects washed ashore.

Sublittoral zone. That part of the benthic environment between the low-tide mark and the edge of the continental shelf (see also *Shallows zone; Continental shelf*).

Submarine canyon. A long, steep-sided valley found at continental margins, usually perpendicular to the coastline. Mud and silt avalanches frequently pour down these canyons from the shallows zone above.

Symbiosis (literally, "living together"). A relationship between two species in which one or both are benefited and neither is harmed. A well-known example is the reef-building corals and the algae (zooxanthellae) embedded in their tissues. The algae consume the wastes and carbon dioxide from the coral animals; the corals provide the algae with a stable base and a supply of nitrogen and phosphorous for photosynthesis (see also *Commensalism; Mutualism*).

Table reef. A small isolated reef rising from the continental shelf, with no lagoon.

Tidal bore. A wave of water that advances up an estuary as the tide rises. It may also be generated by a tsunami.

Tidal marsh. A mud flat; broad muddy or sandy area in an estuary or bay exposed by retreating waters at low tide. The marsh, usually fringed with a dense growth of reeds and sedges, may be dissected by numerous streamlets. The mud flat is often well populated with crabs, worms, and clams, which attract gulls, limpkins, and herons seeking prey. In terms of the food produced by decay of marsh grasses, the tidal marsh is one of the most productive areas in the world.

Tide. The cyclic rising and falling of the sea caused by the mutual gravitational attraction of the Earth, moon, and sun. The moon has the greatest effect (see also *Neap tide; Spring tide*).

Trawl. A large, conical, open-mouthed net dragged along the sea bottom to catch fish and, occasionally, lobsters and other invertebrates. Trawl nets can take upwards of 15 tons of fish in one hour.

Trawler. A vessel that engages in ocean fishing with a trawl net. Modern stern trawlers "shoot" the trawl and haul it back up a stern ramp.

Trenches. Long, narrow, deep depressions in the sea floor with comparatively steep sides. The deepest trenches, more than 10,000 meters below the water surface, were long thought to be devoid of life. Deep-sea dredging and observations from manned submersibles have revealed living organisms in even the deepest trenches.

Trophic level. A stage of nourishment as represented by links in the food chain. Algae, phytoplankton, and kelps are at the lowest level; carnivores, including barracuda, sharks, killer whales, and man, are at the top level.

Tsunami. A protracted high-velocity sea wave produced by a submarine earthquake, volcanic explosion, or landslide. The wave may not be detected at sea, but it assumes gigantic proportions by the time it reaches shore, resulting in great damage to shorefront facilities.

Typhoon. A severe tropical cyclone in the western Pacific Ocean (see also *Cyclone; Hurricane*).

Volcano. A conical landform, often rising out of the sea, with a vent at the summit created by the emission of molten liquids such as lava, steam, hot gases, rock, and cinders. Cone- or dome-shaped volcanoes are temporary features that are eventually destroyed by erosion, especially wave action. Charles Darwin explained the relationship between volcanoes and the coral reefs that create atolls. One of the newest volcanic islands, Surtsey, erupted off Iceland in November, 1963.

Water mass. A body of water in the sea characterized by a particular temperature, salinity, and set of chemical characteristics. Many marine organisms remain within a particular water mass because its physical characteristics are favorable to their life.

Waterspout. A tornado-like wind column, most common over tropical and subtropical waters. It appears as a funnel-shaped column or spout of water reaching from the sea to a cloud and is held erect by a circular wind movement. Waterspouts are not so destructive as tornadoes.

Index

Page numbers in bold face
type indicate photographs